AF335031

# PEREGRINE

# PEREGRINE

EMMA FORD

WITH ORIGINAL PAINTINGS BY ANTONY RHODES
AND PHOTOGRAPHS BY PIERS CAVENDISH

FOURTH ESTATE · *London*

First published in Great Britain in 1993 by
Fourth Estate Limited
289 Westbourne Grove
London W11 2QA



A catalogue record for this book is available from the British Library.

ISBN 1–85702–105–3



Typeset by York House Typographic Ltd
Colour reproduction by Leeds Photolitho Ltd
Printed in Italy by Amilcare Pizzi SpA, Milan

# CONTENTS

Who could ever forget his first encounter with a peregrine? Mine occurred on Thanksgiving Day, 1943, when I was fifteen years old. I was on an outing with a friend, and we were camped by a lake in the foothills of the San Gabriel Mountains above Los Angeles, California. Strolling the shoreline we found several tell-tale carcasses of coots lying about, each headless, well plucked except for the wings, with all the flesh cleanly stripped away from the breastbone and keel. As we paused to examine one, wondering whose work it was, we suddenly heard a loud whistle overhead like the sound of an artillery shell and we looked up just in time to see the stooping form of a falcon as she shot out over the lake and struck down a coot struggling to leave the water. When she turned back to pick up her floating booty, we could see the unmistakable black cheeks and slaty back of the peregrine.

Much of my future life became crystallised in that captivating moment, for I was at once caught up in the mystique of this superlative hunter – I had to experience more, know more about this bird and its way of life, so bold and free and splendid. For fifty years I have followed the fortunes of the peregrine through good times and bad. Fortunately, now, the good times prevail once more, for peregrines can again be seen and heard courting and breeding at many eyries throughout their worldwide range, thanks to the fact that human beings have been willing to alter their behaviour profoundly in order to ensure the survival of this great bird.

It came as a great shock and puzzle to biologists in the post-World War II era when peregrines suddenly and simultaneously began to disappear from established eyries in Britain, Europe, and North America. The story of the effects of organochlorine pesticides such as dieldrin and DDT on peregrine populations is well known: through a combination of effects resulting in both increased mortality and reduced reproductive output, the massive application of these toxic chemicals to the environment was able to accomplish in a few years time what no other form of human persecution had been able to do in centuries. Who would have supposed that the simple laboratory trick of adding a few chlorine atoms to a hydrocarbon molecule could cause so much havoc?

Conservationists, biologists, and falconers on two continents mounted the largest ever effort to save a species from extinction – first by fighting for effective controls on the use of harmful pesticides, followed by habitat protection and management, and by captive propagation and reintroduction. Falconers played an absolutely essential role in the latter activity, and the re-established breeding populations in parts of Germany, Sweden, Canada, eastern United States, the Rocky Mountain states, and California exist today as a tribute to their special knowledge of the peregrine and to their zealous determination that peregrines should not perish in the wild.

Although the peregrine still remains absent from some parts of the former breeding range where it completely disappeared – for example, most of the south coast of England, the forests of northern Germany and Poland, the river banks of the northern Great Plains of North America – the status of wild peregrines is certainly secure for the forseeable future. The several hundred peregrines and other hunting hawks being bred in captivity each year make it possible for falconry to survive as well. As long as enough men and women of goodwill and intelligence remain devoted to its welfare, the peregrine will persist, and 'its dispassionate, brown eyes' will continue to witness our human struggle for an equitable and creative accommodation to the limitations of life on earth. As this century draws to a close it is a most apposite moment to celebrate the survival of the peregrine.

*Tom J. Cade, the Peregrine Fund, Boise, Idaho, USA.*

# ACKNOWLEDGEMENTS

I would like to express my grateful thanks to the following people and organisations who have helped or contributed material: Professor Tom Cade for the foreword; Philip Tose, for making this book possible and Andrew Jamieson for his liaising; Dick Treleaven for his invaluable advice and for acting as specialist reader of the manuscript; Maher M. Al Tajir for his assistance with the manuscript and for access to his reference library; Sheikh Rashid bin Khalifa al Maktoum; the Peregrine Fund Inc.; Dr Clayton White; Ron Hartley; Pete Gill; Graham Butterworth; Patrick Walsh; Stuart Rossell; Peter Blest; Clive Priddle for patient copy-editing; the publishers – Fourth Estate – and Peregrine Group; Fiona Carpenter for her work in designing the book; Penny Spencer for typing the manuscript; my husband Stephen for all his help, enthusiasm and support.

*ILLUSTRATION ACKNOWLEDGEMENTS*

Page 12 Ron Hartley, The Falcon College, Esigodina, Zimbabwe;
Page 28 Garlinda Birkbeck;
Pages 34, 35, 36, 38, 40, 109 from *Falconry and Art* by Christian Antoine de Chamerlat, Sotheby's Publications, Philip Wilson Publishers Ltd, London and M. Rafif, Directeur, ARC Edition Internationale, Courbevoie, Paris;
Page 39 The Tryon Gallery, London and Andrew Miller Mundy for supplying the transparency of the original painting;
Pages 52, 53 RSPB Photographic Library;
Pages 81, 94, 95, 97, 99 Bill Burnham and the Peregrine Fund; the Peregrine Fund;

the Peregrine Fund; the Peregrine Fund; Glen Eitmiller and the Peregrine Fund; Pages 126, 133 Maher M. Al Tajir; Page 142 R. T. Smith and Ardea.

*TEXT ACKNOWLEDGEMENTS*

*The author and the publishers gratefully acknowledge the following for their permission to use copyright material:*

The Peregrine Fund Inc. for *Peregrine Falcon Populations, Their Management and Recovery*, edited by Tom J. Cade, James H. Enderson, Carl G. Thelander and Clayton M. White; HarperCollins, London for *The Falcons of the World* by Tom J. Cade and *The Peregrine* by J. A. Baker; Gresham Books, Surrey for *The Ornithology of Shakespeare* by J. E. Harting; Dick Treleaven for *The Private Life of the Peregrine*; Rogers, Coleridge and White Ltd in association with International Creative Management, Inc. and Severn House Ltd, London for *Peregrine* by William Bayer; C. W. Daniel Company Ltd, 1 Church Path, Saffron Walden, Essex for *Falconry* by Gilbert Blaine; Nick Lyons Books, New York for *A Rage for Falcons* by S. Bodio; HRH Sheikh Zaid bin Sultan Al Nahayan for *Falconry as a Sport*; Weidenfeld and Nicholson, London for *The Hound and the Hawk: The Art of Medieval Hunting* by John Cummins; T. & A. D. Poyser Ltd, Staffordshire for *The Peregrine Falcon* by Derek Ratcliffe; Sierra Club Books, San Francisco for *Peregrine Falcons* by Candace Savage; Debrett's Peerage, London for *A Bird in the Hand* by Roger Upton; Oxford University Press for *English Hawking and Hunting in 'The Boke of St Albans'* by Dame Julians Barnes edited by Rachel Hands; Stanford University Press, California for the translation of *The Art of Falconry* by Frederick II of Hohenstaufen edited by C. A. Wood and F. M. Fyfe; Chasse Publications, Colorado for *Traité de Fauconnerie* by H. Schlegel and A. H. V. de Wulverhorst.

As someone with a passion for peregrines, I am indeed fortunate: I hunt with my own home-bred peregrines and I live in an area populated by wild ones. I wrote this book during the autumn mornings of 1992, while spending my afternoons flying peregrines at red grouse on the picturesque moors of Scotland. Encounters with the wild peregrines which hunt across the same moors are not infrequent, but a special one will linger in my memory.

On the very top of Braco Castle moor, where the views of the surrounding hills stretch into purple infinity, the dog went on point, indicating that a covey of grouse lay ahead. I unhooded my falcon and she left my fist to gain height. When she was in perfect position, and approximately four hundred feet over the point, I noticed a wild peregrine – a tiercel – also circling, about two hundred feet higher again than my falcon. Consumed with curiosity, I called for the dog to flush the grouse and strained my eyes upwards against the evening sun. As the grouse lifted from the heather both peregrines turned to catch the wind and stooped at them. One behind the other, they plummeted down in unison, until my falcon closed in behind the grouse and knocked it into the heather. Before she could regain position, the grouse broke from the heather once more and was pursued out of sight by the wild tiercel. The incident was unusual, as previous encounters between trained peregrines and wild ones had generally resulted in a skirmish. In this instance, however, the tiercel had obviously come over for a look and was so stimulated by the sight of the grouse that he opportunistically decided to join in, in the hope of an easy meal.

My admiration for the peregrine has been shared by many throughout history, who have prized her for her beauty and her unmatchable aerial ability. It is the peregrine's good fortune that she stirs such strong passions in man, for it was man who pulled her back from the brink of extinction in the 1960s, when DDT had threatened to wipe her from the face of the earth.

This book is the result of a chance meeting with a fellow peregrine enthusiast – Philip Tose. Philip's admiration for the peregrine is such that he named his multinational company after her. Peregrine Group is a financial services company with offices throughout Asia, concentrating its activities on the region's rapidly expanding capital markets and economies. With wry humour, Philip likens the swift and decisive way in which his company conducts business on behalf of its clients to the speed and efficiency of the falcon herself. Whilst discussing with Philip some of the fine books and paintings which the peregrine has inspired over the centuries, we decided to attempt to gather together the best examples and aspects of the peregrine in a single volume.

The resulting book is intended to be a celebration of the peregrine, rather than a definitive exposition of her biology and ecology. As such, I dedicate it unreservedly to those individuals around the world who fought for her survival during the DDT crisis and who now rejoice in her future.

# THE PEREGRINE ACROSS THE WORLD

*Man has emerged from the shadows of antiquity with a peregrine on his wrist. Its dispassionate brown eyes, more than those of any other bird, have been witness to the struggle for civilisation, from the squalid tents on the steppes of Asia, thousands of years ago, to the marble halls of European Kings in the seventeenth century.*

Roger Tory Peterson, *Birds over America*

Falco peregrinus minor – *adult tiercel*.



**P**eregrine, *Wanderfalke*, *pilgrimsfalk*, shaheen, duck hawk – whatever name this cosmopolitan falcon travels under, she is celebrated across the globe for her beauty, her speed and her spectacular powers of flight. One of nature's most efficient and deadly hunters, to witness a peregrine in headlong pursuit of prey is to be privileged to watch a monarch of the air. The powerful image which lingers in the mind is one of a steely bolt from the blue – one which we, earthbound, can never hope to emulate.

What is it about this falcon which so inspires those who have been fortunate enough to study her? Perhaps it is the feeling that, for a few precious moments, you are up there with her, soaring high above the ordinary and the mundane, rejoicing in the perfection of physical form which renders her the supreme aerial predator. Although our own species evolved through hunting, over the centuries we have softened and slipped into an easy, lazy existence in which physical imperfections and sloth are no barrier to survival. Perhaps as we look to the sky we feel a stab of envy for this perfect bird, mistress of all she surveys, riding the wind with the world at her wing-tips.

The peregrine is found on every continent except Antarctica. As a breeding bird, she does not occur at very high altitude – she is absent from the Andes and the Himalayas – nor does she favour arid areas, such as the desert regions of Australia and Central Asia. She breeds on many of the oceanic islands, but is not found in New Zealand, the Faeroes or Iceland, or on any of the Bering Islands north of the Aleutians. With these exceptions, the peregrine is a globe-trotter, for the flexibility of her feeding habits, the diversity of her habitat requirements and her exceptional powers of flight have combined to enable her to populate any terrestrial region of her choosing.

The peregrine varies in colour and size, forming some nineteen different races. The biggest and heaviest of the races is *Falco peregrinus pealei*, the celebrated Peale's

falcon, which occurs in its largest form on the Queen Charlotte Islands, off the coast of British Columbia in Canada. The smallest is *F. peregrinus pelegrinoides*, the Barbary falcon, which together with *F. peregrinus babylonicus*, the red-naped shaheen, is considered by some to be a separate species. In between lie many subspecies, varying in colour from very dark, black-capped individuals to pale creamy forms with minimal moustachial stripes. As there is much overlap in the size and colour of neighbouring races, it is difficult to identify with certainty a particular individual unless its geographic origins are known.

Some peregrines are resident, others are migratory – particularly those nesting in northern latitudes. As they are traced south, so the races decline in size, with the smallest and least typical in colour being found in the deserts of the Middle East and North Africa. However, the variations among races are not limited to size, colour

and migratory habits: the breeding behaviour and the style of hunting also vary, and it is certainly this very adaptability which has enabled the peregrine to populate the globe. Those who study her seek to know her in all her forms, striving to determine the elusive trait which stamps a particular long-winged falcon as a peregrine. While the scientific arguments joining and separating the various strains may continue over the coming centuries, the precise taxonomical stamp of what constitutes *falco peregrinus* is likely to remain something of an enigma.

In appearance, a typical adult peregrine is cream on the chest, with vertical black bars on the legs and abdomen. On the back, she is blue-black, with long, tapering wings and a shortish tail, the feathers of which can be fanned out or shut tightly together for streamlining. Her head is dark-capped, like the mask of an executioner, with a polished dark-grey beak, pointed and equipped with a 'tooth' on each side, used to fit into the vertebrae of her prey to assist her to break its neck. Her body feathers form a sleek armour when she is in flight, enabling her to cut through the wind like a knife, assisted by the perfect symmetry of her form. Her feet are proportionally large, with long, slender yellow toes and curved talons, perfectly designed for catching other birds in flight.

Undoubtedly the peregrine's most striking feature are her eyes. Dark brown in colour, they shine with a predator's insouciance, boring into all they survey as if to strip it to the bone. Perhaps surprisingly, the features of the peregrine are also most expressive – when relaxed or contented, her demeanour is softened by the fluffing out of the feathers under the chin and the loosening of the body feathers. Even her piercing glare loses its intensity when, at rest and at peace, she tucks one foot into her feathers and surveys her world with an amiable benevolence.

The male peregrine is always smaller than the female. His proper name of 'tiercel' denotes this, deriving through Old French from the Latin word *'tertius'* or 'third' – the approximate amount by which he is smaller than the female. The female

*Portrait of a Cornish tiercel.*

peregrine is quite simply referred to as the 'falcon'. Confusingly, 'she' is the accepted pronoun when talking of the species in general.

The private life of the peregrine is not easy to observe. Although she is not particularly secretive in her habits, man is quite simply not equipped to keep up with her while she is on a hunting foray. The sight most commonly seen by the casual observer, therefore, frequently amounts to no more than a fleeting vision of her winging across the skies, clearly intent on some distant objective. Rarely, she may pass directly overhead and inspect her watchers briefly, before summarily dismissing them and continuing on her way. Only at the eyrie can she be more easily observed, although the risk of unwitting disturbance at the nest site has caused the authorities of many countries to make it an offence to observe an eyrie from close quarters without a licence.

*When human intruders enter the peregrine's territory she may pass overhead to inspect them briefly.*

Falco peregrinus peregrinus

At rest, the peregrine is easily flushed from her roosting point by a human presence. On the wing, her whereabouts are frequently betrayed by other birds, which panic at her approach, flocks rising and swirling, vocalising in alarm to alert others to her proximity. Thus opportunities to observe her for any length of time away from the nest are a rare treat. At the right time of year, it may be possible to observe courtship-display flights when the pair are intent on each other, but it is most uncommon to witness an actual kill.

Over the centuries, the peregrine has been called by many different names. Some relate to her colour, such as 'blue hawk' or 'grey falcon'; others pertain to her physical attributes, such as the North American name of 'great-footed falcon'. Yet another group refers to her tastes in game – 'duck hawk', again from North America. The Fijian name for the peregrine is '*ganivatu*', which is literally translated as 'duck of the rock'. (A Fijian dictionary in 1850 amusingly defined '*ganivatu*' as 'The name of a very large bird, perhaps fabulous, said to live in holes and eat men.' Ironically, or perhaps in a spirit of misguided self-defence, later Fijians are reputed to have eaten young peregrines taken from the nest.) But by far the greatest number of names – including the name 'peregrine' itself – describe her wandering nature: '*pilgrimsfalk*' in Swedish; or '*pèlerin*' in French; '*falco pellegrino*' or '*peregrino*' in Italian; '*pelegrin*' in Spanish; '*vandrefalk*' in Danish; and '*Wanderfalke*' in German.

Perhaps the most endearing name ever attributed to this long-winged, dark-eyed falcon was a name used in medieval England – 'the fawken gentil'. In this instance, the word 'gentle' must be understood to mean 'honourable' or 'noble', according to the definition of the day. In 1575, in *The Booke of Faulconrie or Hawking*, George Turbervile listed 'seauen kindes of Falcons, and among them all for hir noblesse & hardy courage, & withall the francknesse of hir mettell, I may & do meane to place the Falcon Gentle in chiefe'. A worthy tribute, one might feel, to a bird whose pre-eminence in the eyes of man has undoubtedly stood the test of time.

# THE PEREGRINE IN ART AND LITERATURE

*Then for an evening flight*
*A tiercel gentle which I call, my masters,*
*As he were sent a messenger to the moon,*
*In such a place, flies, as he seems to say*
*See me or see me not.*

Philip Massinger, *The Guardian*, Act 1 Scene 2

For many centuries, the peregrine has proved a great source of inspiration to artists and writers. She has been depicted in portraits and engravings, in tapestries, carvings, sculptures, jewellery and china. Her virtues have been extolled in books, both factual and fiction, in learned papers, on ancient papyri and in present-day literature. Man has attempted to capture the essence of the peregrine in whatever art form he has had available to him through the ages, displaying his admiration for her by his attempts to replicate her beauty and to record her image for posterity.

Shakespeare was obviously a keen ornithologist. He manifests this in a great many of his works, using vivid imagery relating to a wealth of different types of birds, including hawks. It is not surprising that his knowledge of hawks and hawking should have been extensive, for falconry was a popular pastime in the Elizabethan

age. The similes and descriptions which he includes in his works are therefore ones which the audiences of the time would have readily appreciated. Many of the references to falcons, and particularly to falcons 'flying at the brook', would undoubtedly have been intended to depict the peregrine, as in these speeches from Act II, scene i, of *Henry VI Part 2*:

QUEEN MARGARET:

Believe me, lords, for flying at the brook
I saw not better sport these seven years' day;
Yet, by your leave, the wind was very high,
And, ten to one, old Joan had not gone out.

KING HENRY:

But what a point, my lord, your falcon made,
And what a pitch she flew above the rest! –
To see how God in all his creatures works!
Yea, man and birds are fain of climbing high.

DUKE OF SUFFOLK:

No marvel an it like your majesty,
My Lord Protector's hawks do tower so well;
They know their master loves to be aloft,
And bears his thoughts above his falcon's pitch.

DUKE OF GLOSTER:

My lord, 'tis but a base ignoble mind
That mounts no higher than a bird can soar.

CARDINAL BEAUFORT:

I thought as much; he would be above the clouds.

Some of Shakespeare's references specifically use the old names for the peregrine. He correctly distinguishes between the falcon and the tiercel, calling the latter 'tercel gentle'. Within this context, he uses images of the falcon to portray size, strength or superiority. The tiercel is variously spelt 'tercel' or 'tassel', as towards the end of Act II, scene i, of *Romeo and Juliet*:

> . . . O, for a falconer's voice,
> To lure this tassel-gentle back again.

Spenser, in *The Faerie Queene*, also uses the spelling 'tassel'. His descriptions of flights at game are as authoritative as Shakespeare's and, from the style of hawking described, can frequently be assigned to the peregrine, even if she is not mentioned by name:

> As when a cast of Faulcons make their flight,
> At an Herneshaw, that lyes aloft on wing,
> The whyles they strike at him with heedlesse might,
> The warie foule his bill doth backward wring;
> On which the first, whose force her first doth bring,
> Her selfe quite through the bodie doth engore,
> And falleth downe to ground like senselesse thing,
> But th'other, not so swift, as she before,
> Fayles of her souse, and passing by doth hurt no more.

Such an account of a heron goring a peregrine when attacked is borne out by a grisly mounted trio held in the Royal Albert Memorial Museum in Exeter. The group is mounted in the positions in which it was fished out of the River Exe in

Devon. It shows a falcon impaled on the beak of a heron, which is itself held by a tiercel. Apparently the tiercel had the heron so firmly in his grasp that all three, thus locked in mortal combat, fell into the river and drowned. Because the heron is such a dangerous quarry, the taking of a heron with a peregrine used to be considered such a feat that the falcon was entitled to wear a hood bearing the royal colour of purple, as a mark of distinction. (A hood with red side-panels signified a game hawk, and one with green side-panels signified a rook hawk.)

Isaak Walton's *The Compleat Angler* clearly shows that the author is a great admirer of the peregrine. He enumerates each species of hawk used for falconry in pairs, giving first the name of the female and then that of the male:

> The gerfalcon and jerkin,
> The Falcon and tercel-gentle . . .

*Portrait of an immature falcon by Garlinda Birkbeck.*



We can thereby clearly ascertain that by 'falcon' he means a female peregrine, particularly as he denotes her name with a capital:

The element of air is most properly mine, I and my hawks use that most, and it yields us most recreation. It stops not the high soaring of my noble, generous Falcon; in it she ascends to such a height, as the dull eyes of beasts and fish are not able to reach to; their bodies are too gross for such high elevations; in the Air my troops of Hawks soar up on high, and when they are lost in the sight of men, then they attend upon the converse with the Gods; therefore I think my Eagle is so justly styled Jove's servant in ordinary; and that very Falcon, that I am now going to see, deserves no meaner title, for she usually in her flight endangers herself, like the son of Daedalus, to have her wings scorched by the sun's heat, she flies so near it, but her mettle makes her careless of danger; for she then heeds nothing but makes her nimble pinions cut the fluid air, and so makes her highway over the steepest mountains and deepest rivers, and in her glorious career looks with contempt upon those high steeples and magnificent palaces which we adore and wonder at; from which height, I can make her to descend by a word from my mouth, which she both knows and obeys, to accept of meat from my hand, to own me for her Master, to go home with me, and be willing the next day to afford me the like recreation.

In 1967, a book entitled *The Peregrine* was published in Great Britain. This volume, written by J. A. Baker, describes in the most haunting and lyrical language one man's love affair with wild peregrines. As an ardent peregrine-watcher for many years, Baker decided to commit his passion for the peregrine to print, in the form of a daily journal describing his observations and experiences:

To be recognised and accepted by a peregrine, you must wear the same clothes, travel the same way, perform the same actions in the same order. Enter and leave the same fields at the same time each day, soothe the hawk by a ritual of behaviour as invariable as its own. Hood the glare of the eyes, hide the white tremor of the hands, shade the stark reflecting face, assume the stillness of a tree. A peregrine fears nothing he can see clearly and far off. Approach him across open ground with a steady unfaltering movement. Let your shape grow in size but do not alter its outline. Never hide yourself unless concealment is complete. Be alone. Shun the furtive oddity of man, cringe from the hostile eyes of farms. Learn to fear. To share fear is the greatest bond of all. The hunter must become the thing he hunts.

The book was hailed as a masterpiece. The journalist and broadcaster Kenneth Allsop described it as 'both science and poetry. It is crowded with astonishingly close, original observation. The pages dance with image after marvellous image, leaping forward brilliant, direct to the retina.' Such praise is well justified, for the peregrine so inspired Baker that his descriptive powers were enhanced to a pitch where even readers who were unfamiliar with the bird itself could picture the scenes he describes in their mind's eye:

I expected the hawk to drop from the sky, but he came low from inland. He was a skimming black crescent, cutting across the saltings, sending up a cloud of dunlin dense as a storm of bees. He drove up between them, black shark in shoals of silver fish, threshing and plunging. With a sudden stab he was clear of the swirl and was chasing a solitary dunlin up into the sky. The dunlin seemed to come slowly back to the hawk. It passed into his dark outline and did not re-appear. There was no brutality, no violence. The hawk's foot reached out, and gripped, and squeezed, and quenched the dunlin's heart as effortlessly as a man's finger extinguishing an insect. Languidly, easily, the hawk glided down to an elm on the island to plume and eat his prey.

In 1913, Francis Heatherley's *The Peregrine Falcon at the Eyrie* was published by

*Well camouflaged coastal peregrine.*



*Country Life*. It contained the first complete set of photographs ever taken from a hide of a peregrine at the eyrie. Shot in the Scilly Isles, the album includes photographs of eyasses – young birds – in the nest at various ages and the falcon and the tiercel brooding. At the time, this book represented a significant landmark in ornithology, and even today it is still considered a remarkable set of pictures. The text, however, is a trifle confused, as, being in a hide, Heatherley did not see any of the food-passes, and he was not always sure of his identification of the sexes.

In 1923 came the publication of *The Peregrine's Saga* by Henry Williamson, who also wrote the celebrated *Tarka the Otter*. The *Saga* was chiefly famous for its 1945 edition, which included some superb woodcuts by C.F. Tunnicliffe, but overall it never achieved the same public acclaim as *Tarka*. It featured the adventures of Chachek, the Backbreaker, who eventually lost an eye in combat with a heron called Old Nog. Sadly, the book is inaccurate and anthropomorphic, though it was misguidedly hailed at the time as a magnificent insight into nature.

After twenty-five years of peregrine-watching, a Cornishman, Dick Treleaven, wrote a fine monograph: *Peregrine: The Private Life of the Peregrine Falcon*. Approaching his subject in a manner which was both rapt and scientific, Treleaven described the coastal peregrines which hunt over the magnificent Cornish headlands. Of his passion to observe the species, he says:

I cannot explain why my intrusion into the private life of the peregrine is such an uplifting experience. Perhaps because I feel I am sharing, for a few hours, the pleasures of their wild and lonely existence. I enjoy just sitting and watching as they cleave the sky with that immense sense of purpose, the lords of all they survey.

The peregrine has also inspired literary works of a different nature. In 1981 a bizarre novel entitled *Peregrine*, by William Bayer, was published. The story unfolds to reveal an abnormally huge peregrine, trained to kill women in New York. However fantastic the tale may be, the book is well researched and contrived, blending fact with fiction to produce an intricate suspense novel, in the great tradition of the American thriller:

He raised his head as if to catch the sun for warmth and then he saw the falcon in her stoop. What ecstasy! She was falling at an enormous speed. A hundred fifty miles an hour, maybe more, falling, falling, wings swept back, head down pointing at the ground.

She was seconds from her strike and no one had noticed yet. She was making that wonderful half twist in the stoop which was her speciality, the trademark of her attack. A half twist two hundred feet above her victim, final corrections depending upon the wind, lining up her talons with the head, preparing for the close. Yes, the close! It was magnificent! The best one yet – by far the best. When she hit the girl, knocking her off the fountain ledge, it was he, too, who hit her; her talons were his; he was Peregrine.

For centuries, falconers have searched for the right words to describe the peregrine, whom they know more intimately than many of the ornithologists, for they have had the opportunity to experience her temperament at first hand. In the 1930s Gilbert Blaine wrote of the peregrine, 'Its courage, docility, and the brilliance of its performance cannot be rivalled by any bird that flies.'

That aloof calmness of the peregrine's nature which earned her the medieval name of 'falcon gentle' was further analysed by an American writer, Stephen Bodio, in his spirited contemporary work *A Rage for Falcons*:

An adult-caught peregrine will seem shy, but cool and even phlegmatic; she will more likely ignore the falconer than scream at him in rage. It is a little too tempting to wonder if the polite, ladylike composure of the peregrine wasn't as least as important to its popularity among the stiff-upper-lip English falconers as were its stunning stoops at red grouse.

*A. Thorburn, adult peregrine in flight, 1918.*

In contrast, HRH Sheikh Zaid, Ruler of Abu Dhabi and President of the United Arab Emirates, writes in his book *Falconry as a Sport* (1976):

The Arab name *'shaheen'* is derived from the Persian and means 'balance', relating to the fact that this falcon can tolerate neither extreme hunger nor a surfeit of food. The Shaheen falcon is short-tempered and angers quickly, especially when aged. Though it responds to training and discipline it must be handled with gentleness and sympathy. It is said to be 'more fragile than glass'.

In 1988 a well-researched volume by Dr John Cummins was published, entitled *The Hound and the Hawk: The Art of Medieval Hunting*. In this the author admires the writings of Pero Lopez de Ayala, a sixteenth-century Chancellor of Castile, and quotes him frequently on the subject of the peregrine:

For Pero Lopez the greatest glory of falconry is not the imperial eagle, not the gyrfalcon with its kingly robe of ermine, but the peregrine, nebli; smaller, less magnificent, but faster, bolder; lofty in two senses; full of pride: 'I have worked more with the nebli . . . for in truth this is the noblest and best of the birds of prey, the lord and prince of hunting-birds, and whoever may govern and control the nebli may more easily govern the rest.'

In exploring the symbolism of falconry, Cummins came across an anonymous Spanish lyric in which the peregrine is both God and Christ:

> To the flight of a heron
> the peregrine stooped from the sky,
> and, taking her on the wing,
> was caught in a bramble-bush.

High in the mountains
God, the peregrine, came down
to be closed in the womb
of Holy Mary.

The heron screamed so loudly
that *Ecce ancilla* rose to the sky,
and the peregrine stooped to the lure
and was caught in a bramble-bush.

The jesses were long
by which he was caught;
cut from those webs
which Adam and Eve wove.

But the wild heron
took so lowly a flight
that when God stooped from the sky
He was caught in a bramble-bush.

Perhaps the most famous passage ever written about the peregrine was produced by G. H. Thayer in 1904. He describes the American peregrine as:

perhaps the most highly specialised and superlatively well-developed flying organism on our planet today, combining in marvellous degree the highest powers of speed and aerial adroitness with massive, warlike strength. A powerful, wild, majestic, independent bird, living on the choicest of clean, carnal food, plucked fresh from the air or the surface of the waters, rearing its young in the nooks of dangerous mountain cliffs, claiming all the

Woman with Dog and Falcon – *primitive art
from the Netherlands from the Sauvageot
collection. Musée de Louvre, Paris.*



atmosphere as its domain, and fearing neither beast that walks nor bird that flies, it is the very embodiment of noble rapacity and lonely freedom.

On the subject of the peregrine in art and literature, Derek Ratcliffe in his book *The Peregrine Falcon* (1980) expressed his thoughts on the variety of media in which she has been represented:

Many indeed are the purple passages which this bird has evoked. Together with the eagle, it is one of those creatures which humans invest with powerful symbolism and projections. The raptor brings out the anthropomorphism in us, and the peregrine does so especially. It figures variously in folklore, as heraldic insignia, trade emblems and names, public house names and signs, and postage stamp designs. This is among the bird artists' favourite subjects but evidently a difficult one, for few of the many paintings and drawings that I have seen have exactly caught the real essence of the living bird. Some are accurate but lifeless, others are vital but flawed. Two of the ones that I liked best were on inn signs, but perhaps their setting mellowed judgement, so that the comparison is unfair. But I am being over-critical. Bird paintings give pleasure to multitudes and they are one of the best expressions of the human affection for and interest in the other inhabitants of this world.

On the subject of the peregrine in art, Ratcliffe had a point: in the opinion of those who are well acquainted with the live peregrine, few artists have managed to do her justice. Perhaps it is unfair to expect that a single dimension can capture her beauty, or portray the powerful symmetry of her form, yet so many attempts are disappointing that it is tempting to feel that the peregrine has attracted more than her fair share of artistic incompetence. But it is a sensitive subject – to her many devotees, even the best of her artists can, on occasion, produce paintings or facets of paintings which irritate in their imperfections.

Beyond the early Egyptian papyri and depictions of the sky-god Horus, some of which were probably based on the peregrine, the first work of note to include illustrations showing her was Frederick II of Hohenstaufen's *De Arte Venandi cum Avibus*. The original was dictated in 1247. A French edition produced in 1317 was lavishly illustrated, and showed the peregrine undergoing the various stages of training. The prominence of the peregrine within the illustrations accurately conveys her importance within the sport of falconry at the time.

Monarchs were frequently portrayed by their court painters with a favourite falcon on the fist. Hans Holbein (1497–1543), court painter to Henry VIII, produced many fine portraits showing members of the king's court with their hawks, including Robert Chesman, the king's falconer. During the Renaissance period and through to the late eighteenth century, there was a wealth of fine artists, including Christophe Wilhelm Orde (eighteenth century), Alexandre-François Desportes (1661–1743) and Johann Heinrich Tischbein (1751–1829), who produced detailed canvases depicting falcons and the chase.

In the early nineteenth century, Johann Baptist Sonderland (1805–78) portrayed the peregrine in several major works, the best known of which features Sultan, a peregrine from the Loo Hawking Club. The painting is one of a series of portraits immortalising the best of the Club's falcons. Sonderland also spent much time in the hunting-field with the Loo Club, recreating the day's sport in detailed engravings.

Pierre-Louis Dubourg (1815–73), a famous French animal-painter of the nineteenth century, painted two fine portraits in oil of an adult peregrine and a back study of an immature bird. These now hang in the Musée d'Histoire Naturelle in Paris, accompanied by two similar studies of gyrfalcons.

In the late nineteenth century and the early to mid twentieth century, two artists worthy of mention depicted the peregrine in pursuit of quarry. Karl-Wilhelm-Friedrich Bauerle (1831–1912) produced a series of excellent drawings showing the

Sultan *by Johann-Baptiste Sonderland (1805–78).*

Peregrine Falcon Binding to a Shovellor
Duck – *oil painting by Roger Reboussin
(b. 1881). Musée Internationale de la Chasse et
de la Fauconnerie, Château de Gien.*

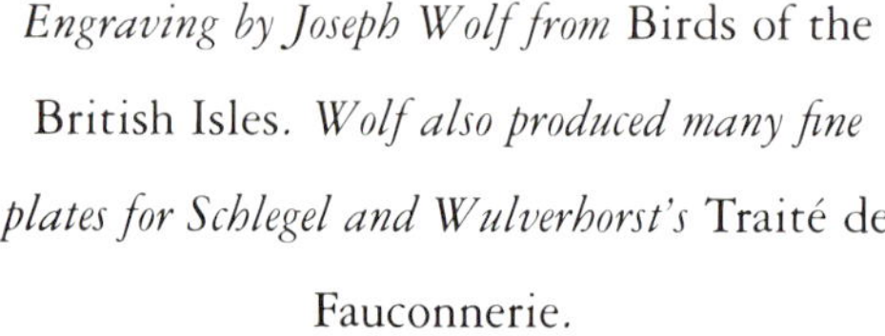

*Engraving by Joseph Wolf from* Birds of the
British Isles. *Wolf also produced many fine
plates for Schlegel and Wulverhorst's* Traité de
Fauconnerie.



peregrine in pursuit of the rook. The accuracy and life encapsulated in these simple sketches clearly demonstrate that the artist had enjoyed many rook-hawking excursions before committing his impressions to paper. In a completely different vein, Roger Reboussin (b. 1881) produced *Falcon Binding to a Shoveller Duck*. In this canvas, he manages to convey something of the speed of the peregrine, and contrasts the colours of the shoveller's rich plumage with a background of muted autumn hues.

Nineteenth-century bird artists found a ready outlet for their work in illustrating the growing number of natural-history texts during the Victorian era, when public interest in natural history increased tremendously and advances in printing techniques opened new horizons. Engraved copper plates were printed on to parchment, and the resulting outlines were then hand-coloured. In this way, limited editions could be produced featuring high-quality colour illustrations. In recent times, many of these nineteenth-century publications have been separated into plates, which are now highly collectable.

The illustrators themselves varied greatly in their familiarity with the subject, and consequently in the quality of their illustrations. The peregrine was included in the vast majority of these natural-history works, and inevitably in any falconry publications of the period. At worst, she was depicted in lifeless, two-dimensional monotones, within which the illustrator showed scant regard for accuracy of feathering or form. At best, the artist managed to convey something of the living bird, with accuracy and sensitivity.

One of the earliest of these major works was John James Audubon's *Birds of America*. The 'elephant'-size portfolio was produced between 1827 and 1838. The illustrations are in vivid colour and are highly stylised. A pair of peregrines is shown depluming a green-winged teal, complete with plucked feathers, blood and gore. Whether or not these plates appeal, it cannot be denied that they are of major importance, representing a milestone in the recording of avian natural history.

Some of the finest illustrative work in the mid nineteenth century came from the hand of the great German draughtsman Joseph Wolf. In peerless detail, he illustrated Schlegel and Wulverhorst's *Traité de Fauconnerie* (1844). These important engravings include an adult female peregrine with intricate feathering detail, perched uncharacteristically in an oak-tree. These plates set the standard for other illustrators, who were later to produce plates for works such as Salvin and Broderick's *Falconry in the British Isles* (1872). This work was illustrated by Broderick himself, who, rather than merely producing a monotypic adult falcon in isolation, also produced plates showing subspecies and colour phases.

Earlier than *Traité de Fauconnerie*, John Gould's *Birds of Europe* (1837) included a

Algerian Falconers, *hunting with peregrines,*
*by Victor Pierre Huguet (1836–1902). Mathaf*
*Gallery, London.*

fine peregrine plate by Edward Lear. Further examples of this type of work are to be found in Gould's *Birds of Britain* (1873), illustrated by Wolf, Thorburn's *Birds of Prey* (1919), and Swann's *Monograph of the Birds of Prey* (1930) with plates by Gronvold.

One of the later artists to be influenced by Joseph Wolf in his studies of birds of prey was George Lodge (1860–1954). He was also a falconer, which enabled him to study the peregrine at close quarters, and to observe her in action. Renowned for his portrayal of atmospheric landscapes, Lodge was able to harmonise his knowledge of the subject with his ability to paint natural backgrounds, to produce sensitive and accurate works in a class of their own. He painted the peregrine in the wild as well as equipped for falconry and in a variety of preening poses which capture the living bird with a precision born of many hours of intimate observation. His work was featured in Lydekker's *Royal Natural History* (1893–4) and his own *Memoirs of an Artist Naturalist* (1946), and was also used to illustrate D. A. Bannerman's *Birds of the British Isles* (1953). Lodge's collaboration with Bannerman started when he was over eighty; 384 of his pictures were chosen from his studio in 1939, but sadly he never lived to see the fruits of his labour in print. The plate of the peregrine in immature plumage was painted when he was ninety-two years old. Later, his falconry plates were used in Jack Mavrogordato's companion volumes *A Hawk for the Bush* (1960) and *A Falcon in the Field* (1966), as well as in numerous other falconry texts.

It is not coincidental that, in recent times, many of the peregrine's best artists are falconers or breeders, and therefore intimately acquainted with her. They have thus been able to capture with pleasing accuracy her moods and body language, depicting her with a three-dimensional precision which is so often lacking in natural-history art. Evidence of the ability of this school of artists is to be found in abundance in *Game Hawking at its Very Best*, by Hal Webster and James Enderson (1988). Within this anthology, plates and sketches by falconers for falconers manage to depict the peregrine in such detail that many of them are almost photographic in their

Black Jess, *painted on the fist of Kim Muir by George Lodge (1860–1954), perhaps the best-known portrait of a peregrine. Reproduced by kind permission of the Tryon Gallery, London.*

Falco peregrinus tundrius *in immature and*

*adult plumage (flying) by Willy Bärr.*



portrayal. John Perry Baumlin's plate of a tiercel on a dove is one such example: the feathering, the expression and the stance are completely typical.

Robert Katona paints in a softer style, producing colour plates which are natural and convincing. In addition, his monochrome sketches of the peregrine bring her to life with a vivid clarity in a medium not often exploited by other peregrine artists.

Mark Waller enjoys outstanding artistic ability. His skill in depicting feathering and detail shows a perfect mastery of his subject. In *Game Hawking at its Very Best* his plate of an immature peregrine chasing waterfowl is a prime example of his work. Other contributors to *Game Hawking at its Very Best* who have produced high-quality plates of the peregrine include James Enderson, Alan Suliber and Don Malik.

It is intriguing that both Renz Waller and Heinz Meng, who made worthy contributions to the early attempts to breed peregrines in captivity, were also talented artists. Renz Waller's sensitive studies of peregrines were used to illustrate his manual on falconry, *Der Wilde Falke ist mein Geselle* (1962). A Swiss contemporary, Willy Bärr, devoted his artistic ability to studies of hawks used for falconry. One of his most celebrated works is a lithograph of tundra falcons, tinted with watercolour. It is, perhaps, inevitable that Willy Bärr was also a keen falconer.

No piece on the peregrine in contemporary art can be considered complete without mentioning David Reid Henry and R. David Digby. Both share an innate knowledge of their subject which has enabled them to produce work which many consider to be the zenith of falconiforme art. Both falconers, they spent time together in Zimbabwe, where they studied falcons in the field and Digby built on artistic techniques which Reid Henry had earlier helped him to master. Digby later painted all the plates in Tom Cade's *The Falcons of the World* (1982).

Among many fine books which have been published in recent years on the subject of falcons and falconry, one of particular interest to falconers worldwide is Mark Allen's *Falconry in Arabia* (1980). This book is illustrated by Mary-Clare Critchley-

Salmonson, who portrays her subjects with realism and precision. Among the plates, she has included an original study of peregrine feathers and the special technique of repairing them, known to falconers as 'imping'.

In concluding this brief list of artists who have been inspired to paint the peregrine in modern times, it is of course only fitting to mention Antony Rhodes, himself a falconer, who has produced the majority of the paintings in this book. Antony's passion for the peregrine is such that he spends many hours not just in observing his own falcons but in travelling to see those of other falconers. His talent for producing a painting which is not merely identifiable as a peregrine but is furthermore unmistakably a certain individual is legendary. This rare quality has advanced Antony to the realms of the great avian artists whose work has immortalised the peregrine.

*Wild peregrines love the wind, as otters love water. Only within it do they truly live.*

J. A. Baker, *The Peregrine*

*Peregrines prefer running water for bathing,*

*preferably shielded by vegetation.*

From seventy-six degrees north to fifty-five degrees south, peregrines live and breed in habitats as diverse as the Arctic tundra, temperate forest zones and coastal inlets. Their basic requirements are open ground to hunt, a plentiful supply of prey, and suitable natural nest sites. They even manage to nest in urban environments, using towering office buildings in place of high cliffs or crags. Many of the nest sites have been used for centuries, by generation after generation of peregrines, in a phenomenon known as 'ecological magnetism'.

Many scientists claim that the area in which young peregrines hatch and subsequently fledge exerts a strong pull over them for the rest of their life. Young birds may, they say, 'peregrinate' until they mature for breeding, but they will then try to return close to the area of their birth to find a territory of their own. Others dispute this theory, saying that young peregrines will drift as far as is necessary until they find a good food supply – usually close to a pair which is successful. Recent research shows that tiercels will typically find a territory within about 60 miles of their natal area, whereas falcons will cover up to 125 miles seeking a mate. Peregrines normally pair for life. When one of the pair meets with an accident, the remaining bird will select another mate from the surplus stock of peregrines of breeding age who have failed to secure a territory. The new pair will then stay within the territory of the survivor. It has even been rather unromantically mooted that the nest site itself may well be the attraction for the pair, rather than each other. This philopatry – loyalty to natal areas – has enabled detailed observation of the private life of the peregrine.

Peregrines live up to twenty years, and remain reproductively active throughout. The first couple of years, before the individual reaches maturity, are usually spent largely in isolation, although immature peregrines may form non-breeding pairs. A peregrine generally reaches maturity at the age of two, although it may be three or four before it breeds. The highest mortality rate – estimated at as high as

71 per cent – occurs during the first year, when peregrines are in their immature plumage and their inexpertise at hunting and bad weather conditions may conspire against them. Deaths thereafter can be attributed to a number of factors, including parasites and disease, poisoning, persecution and starvation.

During their immature year, young peregrines will seek an area where there is a stable food supply and, to find this, even the non-migratory races may become partially migrant as they seek autumn and winter hunting-grounds. In the following spring, immature birds which have been hunting on low ground for the winter will move back to higher ground, driven by the urge to seek nesting territories. An immature peregrine will occasionally pair with an adult. Immature falcons have been seen on eggs and even feeding young while paired with adult tiercels, and one immature tiercel has been seen feeding chicks. It is likely that these occurrences arose because of the death of one of the breeding pair, for it is well documented that during the breeding season a bereaved peregrine can replace its mate from the surplus population within twenty-four hours, and that the new mate will take over whatever duties are required – including incubation, brooding or feeding chicks.

When a young peregrine is not hunting and feeding, she will spend her time patrolling, roosting, preening and bathing. Bathing is important to a peregrine to keep the feathers in good order, so she will seek out a shallow pool, wade in up to chest height and wallow, flicking her wings and tail to shower herself with droplets and dipping with her wings alternately, to scoop water on to her back. When she has finished, she will hop on to a nearby rock or perching-point, where she will 'hang out to dry' while preening her feathers back into place. Peregrines have excellent waterproofing, particularly when they are in adult plumage. Peale's peregrines, which live in the high humidity of the Queen Charlotte Islands, have a waterproofing powder on the feathers which is unique among peregrines and is highly effective against the hazards posed by the island's damp climate.

*Meticulous preening is an important part of a peregrine's daily routine.*

The preening ritual is one to which a peregrine will devote time and meticulous attention, for feather care is of paramount importance to an avian predator. At the top of her tail she has a preen gland, which contains a waterproofing oil. By nibbling gently at the gland, she can extract the oil, which she then wipes on to her feathers with her beak. Each flight feather is individually drawn through the beak, and the twelve tail feathers are flexed in arcs around the body as she preens them from top to tip. The body feathers are held erect from the torso as she works through them, closing her eyes while she preens the ones closest to her head. She uses her feet to scratch her head feathers, and usually finishes the performance with a rouse – when all the body feathers stand on end, and she shakes them back into place. She will also rouse immediately before flying.

The feather quality of all the immature peregrines is softer and more flexible than that of the adult falcon. This is possibly to reduce the risk of an immature bird breaking a feather through adolescent clumsiness.

Sometimes a peregrine will stretch her wings while at rest. In an action called 'mantling', she will drop each wing in turn, stretching it down below her body. The corresponding leg is flexed downwards at the same time. After repeating the action on the other side, she will 'warble', or stretch both wings upwards simultaneously, in a hitched position, straightening to enable her to complete the ritual with a few rapid wing-beats.

In the early spring, when the urge to breed overtakes a peregrine which has reached sexual maturity, she will begin to seek a suitable territory which is not already occupied by an established pair. The actual pair-bonding process begins as one individual joins another which is already in possession of a suitable breeding territory. After some vocal persuasion from the resident bird, the visitor decides whether to stay or to move on. The first sign that a pair bond has been formed is generally the perching together of the two individuals – some distance apart at first,

OPPOSITE

*The falcon stretches her wings after preening.*

47

*A tiercel performs a fast fly-by — one of the ritual manoeuvres demonstrated during courtship display.*

and then closer and closer until they roost side by side. Once this initial bond has been forged, the pair go hunting together, first merely in the proximity of each other, then joining forces to target their attentions on the same victim. Aided by the sleights and feints of the stoops – the swoops on to the prey – of one of the pair, the other finds it easier to achieve an easy kill. It is interesting that the falcon always feeds first, regardless of which of the pair achieved the kill.

Nobody who has observed a pair of peregrines for any length of time can deny that they fly for the sheer pleasure of it. Their obvious revelling in their powers of flight is never more apparent than during courtship display. This is frequently triggered by driving away an intruder from within the pair's territory. After it has been chased away, the tiercel usually decides it is time to show off, and what follows is surely the most spectacular aerial display in the natural world. It has perhaps been best described by J. A. Hagar, who studied the peregrine in Massachusetts:

Again and again the tiercel started well to leeward and came along the cliff against the wind, diving, plunging, saw-toothing, rolling over and over, darting hither and yon like an autumn leaf until finally he would swoop up into the full current of air and be borne off on the gale to do it all over again. At length he tired of this, and, soaring in narrow circles without any movement of his wings other than a constant small adjustment of their planes, he rose to a position 500 ft or so above the mountain of the cliff. Nosing over suddenly, he flicked his wings rapidly 15 or 20 times and fell like a thunderbolt. Wings half closed now, he shot down past the north end of the cliff, described three successive vertical loop-the-loops across its face, turning completely upside down at the top of each loop, and roared out over our heads with the wind rushing through his wings like ripping canvas. Against the background of the cliff his terrific speed was much more apparent than it would have been in the open sky. The sheer excitement of watching such a performance was tremendous, we felt a strong impulse to stand and cheer.

Ledge display in peregrines takes the form of food-passing and the taking-up of a number of distinctive postures, performed by both sexes in the immediate proximity of the intended nest site. A great deal of head-bowing takes place, sometimes followed by a touching of beaks.

The tiercel takes over all the hunting henceforward, until the time when the falcon has completed her duties of laying, incubating and brooding. When taking food to the falcon, the tiercel may leave it on the ledge if he is nervous, or he may tiptoe towards her with it in a most self-effacing manner. In the early stages, the female will usually snatch it from him, but she will begin to accept the tiercel's offerings more gracefully when the food-passing has become an established routine for some days. The transfers of food are accompanied by wailing and kecking calls. As a variation,

*Eyries are selected on sheltered cliffs, out of the prevailing wind.*

the tiercel will sometimes pass food to the falcon in mid-flight, dropping the dead bird for her to catch, or passing it to her from foot to foot.

The falcon will usually begin to solicit for copulation a day or two before the tiercel actually mounts her. She will lower her head and raise her feathers, sometimes fanning her tail and wailing at him to indicate that she is ready. Before copulation, the tiercel generally executes a particularly impressive aerobatic manoeuvre from his repertoire and alights directly on the back of the falcon, bending his talons inwards to avoid injuring her while mating takes place. Alternatively, he will alight on the ledge beside her and hitch his wings, walking towards her with a stiff-legged gait. Actual copulation lasts for an average of eight to ten seconds, with the tiercel balancing himself with short wing-beats. Sometimes mating is silent; sometimes it is accompanied by a lot of noise.

Copulation can start several weeks before the first egg is laid, but will increase in intensity a week before laying. It will usually continue until the penultimate egg is laid. The eggs are laid in a scrape on a ledge. Like other falcons, peregrines do not build nests but use their feet to dig a shallow hollow out of turf or loose scree on the cliffs, and use their beaks to build up the side of the scrape. Frequently, several scrapes are made during courtship before the falcon selects her favourite, after turning round and round in it, nestling down into it and trying it for size. The eggs are a mottled brown colour and are produced at intervals of two to three days. Clutch size averages three or four eggs. The falcon will not start to incubate until the penultimate egg has been laid. This ensures that, when they hatch, there is no more than a two-day gap between chicks. The tiercel likes to share part of the incubation, but he is largely kept occupied by hunting for both himself and the falcon during this time. At night, only the falcon will incubate, for it is colder and therefore more important that the eggs are thoroughly covered.

The incubation period for the peregrine varies between thirty and thirty-three

*During the hottest part of the day the falcon*

*shades the eyasses from the sun.*

days, and the eggs hatch in late spring. As the time approaches for a chick to break out, it cheeps inside the shell and the falcon responds by chupping to it, encouraging it to break out. From the time when the chick first makes a crack in the shell to the time when it finally emerges can be as long as seventy-two hours or more. After hatching, the chick is exhausted, and damp from the moisture inside the egg. At this stage it is very vulnerable to chilling, so the falcon broods it tightly. After some hours, it dries off and begins to fluff up, until white down covers its body like thistledown. At this stage it measures approximately two and a quarter inches in length and scarcely has the strength to lift its head. After some hours, it accepts its first feed from the falcon, who tears up the food brought in by the tiercel into minute pieces and offers them tenderly to the chick from the tip of her beak, moistened by secretions from her nares or nostrils.

*The falcon is stimulated to feed by the begging*

*calls of the eyasses.*

When the young are hatched, the tiercel rarely broods them. Being smaller than the female, he has trouble covering the brood of chicks, and he also has to hunt for the family – a task which becomes increasingly arduous as the chicks develop.

The chicks grow rapidly. When they first hatch, their eyes are closed, and they remain so until the chicks are about four days old. The first soft covering of white down becomes increasingly sparse as the chicks grow, until it is replaced by a second coat when they are approximately ten days old. As the chicks grow bigger, they are brooded less tightly by the falcon during the day, and they begin to move about on the ledge. Their first proper feathers appear after about three weeks, and they assume a spiky look as the quills protrude through the down. By this stage the chicks are eating voraciously, having accepted progressively larger pieces of meat – complete with bones and feathers – from the parents, until they are snatching whole food items

and eating them unaided. The arrival of food signals a crowding forward of the eyasses, who may have a tug of war until one secures the offering and shuffles off to a quiet area of the ledge. It then proceeds to drop its wings and mantle over the booty, although eventual possession is seldom contested by siblings.

At the age of five weeks, the young peregrines are the same size as the adults. Most of the down has disappeared, and the spiky look has been replaced by a sleek coat of immature feathers. The feathering of the young peregrine is totally different to that of the adults. On the back, the young birds are brownish, whereas the adults are blue or black. On the chest, the immature bird has vertical streaks in place of the distinctive horizontal barring of the adult. The immature bird's tail is longer, to aid manoeuvrability while she is learning to fly. Throughout her first year, she will be cloaked in the mantle of the beginner, until she moults the following spring and assumes her adult plumage. The eyasses make their first flight in early summer, some five to six weeks after hatching. Tiercels develop more rapidly than falcons and are usually the first to try their wings. Some observers have noted the parents using a number of different ruses to persuade the eyasses to make their first sorties from the nest and the surrounding ledges. These have included hovering with food tantalisingly just a few feet from the ledge, and deliberately cutting down the rations supplied to the eyasses to increase their hunger and thus their desire to take to the wing.

When an eyass finally takes the plunge, its first flight will take it to a nearby ledge, where, apparently lacking the coordination to return to the natal ledge, it will stay for a while, calling to its parents for food. For the first few days, short flights are interspersed with feeding from carcasses delivered by the pair and with periods of rest, during which the eyass will lie down on a ledge, dozing or watching the skies idly. During this period, the eyasses call plaintively for food, vibrating their feathers and wings to attract the adults' attention. When the parents fail to respond quickly enough, the eyasses fly after them, initiating aerial food-passes.

*The brown back panel of an immature peregrine.*

Food-passes between parents and eyasses occur in two ways. Initially, the prey is usually passed from foot to foot. The eyasses are extremely clumsy during their early attempts to secure the food in this manner, often dropping it and causing the adults to have to stoop rapidly to catch it before another attempt is made. Alternatively, the adult peregrine may climb above an eyass in flight and drop the prey for it to catch. Gradually the eyasses become more adept in receiving both types of food-pass, and the quantity of food delivered directly to the roosting ledges is reduced accordingly.

The eyasses embark upon a series of 'getting-ready games', spending the long days of summer practising for the time when they will begin hunting for themselves. In between dozing and preening, they chase each other playfully and begin to irritate other birds on the wing by buzzing them. Sometimes these passers-by are much bigger than the eyasses themselves, and include other predators such as buzzards and carrion crows. Target practice frequently includes stooping at vegetation – whisking the heads off flowers and long grasses. When they are more confident and competent, they begin to make serious attempts to catch passing birds. If these are strong fliers, such as pigeons, the eyass will be outflown and therefore unsuccessful, so it will frequently have to content itself with catching insects until its mastery of the air, and of its own physiology, increases.

As the vast majority of the immature birds' early attempts to catch their own prey are unsuccessful, the parents have to help the eyasses in a variety of ways, including dropping live birds to them, shepherding prey in their direction, or tiring a potential victim by harassing it before leaving the eyass to finish the job. Slowly, some weeks after making their first flight, the eyasses learn to become independent. Once they are hunting for themselves, the eyasses will nevertheless remain in the territory of the parents for many weeks. Finally, in the autumn, driven by some internal instinct, they move on.

The siblings will disperse, seeking hunting-territories for the winter which are not already occupied by established peregrines. The winter movements of both adult and immature peregrines are fairly hard to trace – some pairs remain close to their breeding territories, while others disperse and, like the immature birds, seek new hunting-territories. In countries which have severe winters, the adults often stay within their summer territories until the bad weather approaches, but then move to territories on lower ground, where the hunting will be easier and the weather less inclement.

The immature birds face their biggest challenge at this time, for the winter claims many victims as the weaker or unlucky die of starvation. They have much to learn – they may also have to change their diet to correspond to the available prey species – and they will need more food to fuel them in a colder climate. Prolonged heavy rain, high winds and snowfall make hunting extremely difficult for the immature birds, and many perish in direct consequence of protracted bad weather. Once past their first winter, however, mortality is significantly reduced and the survivors can wander freely over the coming years, as befits their name, until the urge to seek a mate and a breeding territory overtakes them, whereupon the life cycle of the peregrine begins once more.

# THE PEREGRINE HUNTING

*While they have a special taste for pigeons, their menu generally reads like a field guide to the small and medium-sized birds of the locality. Peregrines' spectacular proficiency at this challenging mode of life made them the awe and envy of human hunters.*

Candace Savage, *Peregrine Falcons*

eregrines are ideally suited to their role as a killer of birds. They combine speed, power and the ability to be enormously versatile. In Central Europe, 210 species have been recorded as falling victim to the peregrine. These vary from birds weighing no more than a third of an ounce to geese weighing over four pounds. It is rare to find instances of peregrines taking mammals, although *Falco peregrinus tundrius* sometimes take lemmings and voles when these tundra rodents reach 'peak' levels in certain years. Fish, bats, lizards and amphibians have also been recorded as falling to the peregrine's talon.

Individual pairs frequently develop a taste for a certain species, and prey upon it to a degree which is disproportionate in relation to the variety of potential prey species available within their hunting-territory. This is usually either because the species in question is easily distinguished by plumage colouration – for example, the North

*A peregrine will hunt on the ground for lizards*
*and amphibians to supplement her predominantly*
*avian diet.*

American red-winged blackbird, or because it displays a conspicuous flight pattern, such as the teal.

The preferred prey are species which fly individually and are an optimal size for catching for both falcon and tiercel. Pigeons and doves fit the bill admirably and are preferred prey worldwide, wherever they occur in quantity within the range of a peregrine subspecies. Although the difference in size between the sexes would seem to imply that falcon and tiercel could exploit separate niches in terms of diet, in reality they generally prey on the same species, although occasionally the tiercel may bring a smaller food item to the nest when chicks are small. The tiercel does the larger share of the hunting when the falcon is brooding, but, when she is free from this duty, the falcon will take over the majority of the hunting while the tiercel guards the eyrie.

*A tiercel meticulously plucking a feral pigeon.*
*Pigeon is universally the preferred prey of*
*the peregrine.*

Peregrines adopt many different styles of hunting to suit the task at hand. They generally hunt early in the morning or in the evening – except when they are feeding young, when they must kill more often. It is extremely difficult to observe a kill made by a wild peregrine, but easier to observe the tactics adopted. Although the combination of these tactics will be varied to suit the habitat, the weather and the species of prey, it is possible loosely to categorise the most often used hunting-patterns.

A great many pursuits are preceded by 'still-hunting'. The peregrine sits on a vantage point and sights prey, hoping to pick out an individual to chase which is at a disadvantage. Once one has been sighted, the peregrine gains additional altitude, if necessary, before stooping vertically at the prey. She will nearly always aim to bind to her quarry in mid-air, for if she strikes it and knocks it down she may lose it. However, over very open ground she may level out behind and strike with the back talons, often stunning the quarry, or ripping it open. Occasionally the strike will kill the victim stone-dead.

If the initial stoop fails to yield immediate success, the peregrine will turn on her tail and climb steeply before stooping again, aiming to bind either by closing in behind the quarry or, more commonly, by rolling under it and snatching it in her feet. If necessary, the stoop will be repeated several times until capture is successful or the peregrine is beaten. Having bound to the quarry, the peregrine may sever its neck by biting through the vertebrae while still in mid-air.

In an alternative hunting technique, the peregrine will 'ring up' to get above high-flying quarry and perform a series of shallow stoops, or she will tail-chase it and ride it down to the ground. This style of flight is again preceded by still-hunting, but in this case the peregrine is looking not below her but above her, making full use of her superior eyesight, which is reputed to be up to eight times as acute as human vision.

*The peregrine will select a vantage point from*
*which to still hunt, making use of her*
*phenomenal eyesight.*

*A peregrine takes off from a vantage point to*

*intercept prey.*

When she has spied a suitable individual, she takes off and proceeds to climb upwards in tight circles, to gain height. As soon as it has seen her, the quarry tries to evade capture by outflying her, and the battle for altitude is on – some flights may reach over 3,000 feet. This is an excellent hunting technique for intercepting the flight-lines of migratory birds.

Some hunts are initiated while on the wing, as the peregrine soars high or circles, awaiting the appearance of potential prey. This style of hunting also allows the peregrine to patrol her territory. From this position, she can either pump across the sky into a dominant position and stoop, or she can 'ring up' to take a high-flyer which passes overhead. It is unusual for her to hunt in this manner on a clear day, when she would be easily visible, but on cloudy days these tactics are highly effective.

When necessary, peregrines are very efficient at tackling a flock of birds. They will

*At the bottom of the stoop the peregrine levels out
to close in behind the prey.*

attempt to single out an individual – usually the wingman – by splitting it off from the rest of the bunch. Pursuing the flock, the peregrine flies above or below them, making shallow stoops or feigning upward climbs through the mass, trying to get the flock to split. If she is successful in her aim to separate an individual, she will cling tenaciously to it.

Pigeons and doves are only taken easily if surprised. Once they have seen the peregrine, she may have to stoop many times before she can catch one.

Rarely, peregrines have been recorded as hunting in groups. R. Meinertzhagen reported:

In February 1910 on a sheet of water near Meerut I saw eleven peregrines harrying ducks who were loth to leave the surface. We were there to shoot, the falcons were there to hunt. The

**ABOVE LEFT**
*A peregrine patrols her territory, constantly on
the lookout for intruders.*

ducks were loth to take to the wing but were forced to do so by beaters. We co-operated in the ducks' discomfiture. It was a beautiful sight seeing the falcons stooping in all directions and paying scant regard to the fusillade. I counted five strikes all over water all recovered by the falcons.

Although there are many variations within the styles of hunting adopted by peregrines, a distinct style known as 'low-level attack' has been recognised. This is frequently related to the geography of a pair or individual. It is favoured by coastal and island peregrines – particularly Peale's falcons – who skim low over the sea to surprise birds swimming on the surface, but it is also used to a lesser degree by some inland peregrines, who utilise the contours of the terrain to take their prey unawares.

Much has been written about the speeds which peregrines are capable of reaching in a stoop. While it is generally agreed that a peregrine averages about forty to fifty-five miles per hour while cruising in level flight, claims and counter-claims for the maximum speed she can reach in a stoop have coloured her written history for many years. The highest actual recorded speed was timed by Hangte (1968) at 217 m.p.h., at an angle of descent of forty-five degrees.

One might imagine that, with such powers of flight, the peregrine's success rate at quarry would be bordering on 100 per cent. However, the truth is quite different, and is also very hard to express as a percentage. This is because, after much study, it appears that the attacks launched by peregrines on prey must be divided into two categories – 'high-intensity' hunts and 'low-intensity' hunts. These categories were identified by Dick Treleaven, after many hours watching the Cornish peregrines. The former occur when a peregrine needs to achieve a kill; she then drives home her attack and shows dogged persistence. The latter are clearly demonstrated when a peregrine initiates the hunt in one of the usual manners, catches up with her prey, but fails to press home the attack.

A peregrine may hit and leave prey if she has been stimulated to hunt merely by the sight of a sick or disadvantaged bird, rather than by the desire for food. Also, it is likely that low-intensity hunts serve as training flights, to hone the peregrine's skills and to keep her fit. Overall, a peregrine spends very little time actually hunting – even with chicks to feed. According to the observations made by Dick Treleaven, she is capable of spotting a fast-flying pigeon at a distance of up to two miles and can secure it within three minutes. The duration of some hunting flights is often far shorter than this, however. If the peregrine did not indulge in rehearsals, she would be likely to lose her edge.

These two styles of hunting muddy the waters when one tries to work out an average hunt-to-kill ratio – as do the immature birds, who, in comparison to the adults, are markedly inferior in their hunting ability. Indeed, many immature birds fail to survive their first winter because of their inexpertise, and even one-year-old peregrines frequently appear still to have much to learn. Dick Treleaven reckons that, overall, one in four flights is successful. It is certainly untrue to say that a high-intensity hunt always ends in a kill, but, as identification of a high-intensity hunt is a matter of opinion on the part of the observer, it seems unlikely that a true measure of the peregrine's efficiency will ever be made. Aesthetically, it is tempting to feel that her powers of flight are such that she will always achieve a kill if she really wants a victim. Scientifically, however, admiration must be tempered by the facts available.

Cooperative hunting by a pair occurs from the outset of the breeding season and reinforces the pair bond. Dick Treleaven described a hunt which perfectly illustrates the advantages enjoyed by a pair hunting together:

Both birds flew off together coming straight towards us climbing fast – the falcon leading. They passed right over our heads travelling like rockets. The tiercel swung away to the right and made a complete circle going some hundred and fifty feet higher than the falcon.

*A method of hunting termed 'low-level attack' is used over water by peregrines, particularly Peale's.*

Seconds later the falcon turned and came back in our direction in a long stoop. At that moment a party of eight or more pigeons swept past us, low down heading up the coast. The falcon rapidly overhauled the pigeons, and dived through the middle of them causing them to disperse in all directions. Two flew inland up the valley closely pursued by the falcon. She abandoned them and peeled away to the seaward side where two more pigeons were trying to re-orientate themselves and make for the cover of the cliff-face. The falcon saw her chance to head them off and put in a long curving stoop, which brought her up directly under the left-hand bird. For an instant they appeared to touch, the pigeon shot upwards as if fired from a gun. The tiercel was up above waiting. As the pigeon turned, he closed in fast and gathered it up in his talons without checking his flight. A perfectly co-ordinated attack – the falcon had steered the quarry into the tiercel's clutches. He carried the kill back to the main cliff where the falcon alighted beside him and carried it off.

In conclusion, it appears that the key to the peregrine's success as a hunter must lie in the catholicism of her tastes, her speed in the stoop – which is reputedly unsurpassed by any other living creature – and her opportunistic habits. The combination of these assets results in a bird whose reputation for pre-eminence as a predator is generally considered wholly justifiable by those privileged to see her in action.

# THE PEREGRINE AND MAN

*And there at last was the marvellous bird in life, cruising up and down for me to admire, and filling the dale with its strident chatter. I was at one with the gods.*

Derek Ratcliffe, *The Peregrine Falcon*

During the history of man's somewhat chequered relationship with the peregrine, she has represented both friend and foe: in captivity she has brought food for the table, while in the wild she has sometimes appeared on the menu. Both in her persecution and in her protection, man has gone to enormous lengths to succeed.

The story of the persecution of the peregrine spans many centuries and many continents. Even in the egg the peregrine is not safe, for the eggs themselves are so beautifully and variously marked that egg-collectors worldwide vie to include clutch after clutch in their collections. In Britain, the craze began in the Victorian era. From its early beginnings as a hobby for the Victorian gentleman, egg-collecting grew to the stage where ornithology and oology were virtually synonymous. Raptor eggs in general were highly prized, but those of the peregrine – which by this time was becoming rarer as a result of other forms of persecution – were particularly so. Even her cliff nesting habits were no protection: they simply rendered the acquisition of the eggs more of a challenge, as collectors armed themselves with ropes and climbing gear and made daring raids over the edge.

Egg-collecting was most popular in the south of England, and collectors vied to be the first to reach certain sites. By the middle of the twentieth century, the peregrine population in southern England had been decimated. The Natural History Department of the British Museum boasts many nineteenth- and early twentieth-century egg collections and associated memorabilia, including a photograph of a collector and dealer sitting on a Dorset cliff with a pyramid of sixty-four blown peregrines eggs, harvested in 1928. It is ironic that the men who perpetrated such crimes against nature in general and against the peregrine in particular were frequently the very men who were responsible for substantially increasing our knowledge of the peregrine in her natural state. As the producers of the coveted eggs, the birds themselves were admired and studied by the collectors, many of whom looked upon themselves as naturalists and kept copious notes on nest sites and nesting habits.

While egg-collecting had a serious effect on the peregrine population in southern England, it must be admitted that overall, in terms of Europe and North America, it had little impact. In fact, the plundering of peregrine nests has provided useful historical information on her range, for the data cards which oologists complete when they take eggs have enabled naturalists to glean much information about the population dynamics in certain regions throughout Europe and the United States. Sadly, egg-collecting is a die-hard activity, and the collecting of peregrine eggs – now illegal in many countries – continues today as, spurred on both by their obsession with the hobby and by their desire to outwit the protectionists, oologists furtively add to their collections.

In the nineteenth century, those interested in game preservation began to wage war on the peregrine in Europe as it became fashionable to blame her for all manner of ills which beset the shooting community. She started to appear in vermin lists on the great estates, and keepers were encouraged to kill her whenever and wherever they found her. Bounties were paid for her destruction, in the alleged interests of preserving such species as the red grouse on the moorlands of Scotland and the north of England. As early as 1780, the bounty on an individual peregrine in Scotland was 2s 6d, and the bounty paid for the destruction of a nest was 10s 6d. Over the years, horrifying numbers were killed as part of an indiscriminate slaughter of raptors.

The gamekeeper's war against the peregrine was pursued throughout the nineteenth and into the early twentieth century. In Britain in 1890, a keeper on the Isle of Skye was quoted as saying, 'The falcon is a real sportsman; I just grudge him the maintenance of his family.' Systematic attempts to wipe out the peregrine by shooting adults and smashing eggs or killing young in the nest were backed up by trigger-happy shooters who took pot-shots at any peregrine unwise enough to put in an appearance, and by farmers who feared for their domestic poultry.

The collecting of mounted specimens became extremely popular, providing an outlet for the carcasses and yielding substantial financial inducements from the dealers, each of whom was supplied by his own circle of keepers. Derek Ratcliffe quotes John Harvie Brown, writing in 1865:

About the end of April or the beginning of May the peregrine 'crop' commences to be gathered, judging from the numbers annually sent in at that time for preservation to Inverness. From the numbers that are sent in to be stuffed, it is a wonder the bird holds its own as well as it does.

The desire for mounted specimens certainly resulted in the deaths of more peregrines than would otherwise have been shot in the interests of game preservation, and again it is sad to note that this now illegal harvest for the procurement of skins undoubtedly still occurs in Europe, although on a much smaller scale than in the past.

Falconers traditionally took their hawks from the wild – as eyasses, taken from the eyrie; as passagers, hawks trapped during their first year; or as haggards, hawks trapped in their adult plumage. When this became illegal in many countries, under the terms of various wild-bird protection acts, the temptation to procure peregrines by illicit means proved irresistible to a burgeoning group of hawk-dealers. These unscrupulous individuals plundered eyries and trapped peregrines, building up an extremely lucrative trade by selling them to falconers. The ruses used to cover this illegal harvest were complex, and frequently very hard to trace. Thus, for many years, falconers were highly unpopular with protectionists, who felt that the damage wreaked upon wild stocks as a consequence of supplying falconers was as significant as that caused by egg-collectors, gamekeepers and taxidermists.

Among the peregrine's many enemies, there is yet another group which exists in a

state of total belligerency against her. Pigeon-fanciers swiftly discovered that the peregrine fed enthusiastically on their homing pigeons, which, following set flight-lines, were easy meat for a hunting falcon. Hostilities began in the mid nineteenth century, when pigeon-racing began to attract a growing band of enthusiasts. They could easily prove their case against the peregrine, for during the breeding season the rings of their charges were found in considerable numbers in the peregrine's eyries. Fanciers took matters into their own hands up until the time when they were legally prevented from doing so. Thereafter some illegal persecution continued, but, unlike the gamekeepers, the fanciers' preferred route was to appeal to the government for the right to exercise control of the peregrine wherever she was proved to be doing damage to their stocks.

For a short period in Britain, the fanciers had their wish and a full-blooded

extermination campaign was again launched against the peregrine. The aim, however, was not to protect racing pigeons but to prevent the destruction of message-carrying pigeons during the Second World War. Under authorisation from the Secretary of State for Air, hundreds of peregrines and eggs were destroyed in certain specified areas.

After the war, the peregrine again enjoyed official protection, but her numbers were soon to be depleted across several continents by the most deadly enemy she had yet to encounter: man-made pesticides. Ironically, it was the pigeon-racing fraternity who first drew attention to the plight of the peregrine. In 1960 in Britain, Ivor George, a leading light of the South Wales fanciers, appeared on television to plead for some method of controlling the peregrine, which he claimed was decimating fanciers' lofts. The Home Office, in need of concrete facts and figures, asked the Nature Conservancy Council for information. It, in turn, contracted the British Trust for Ornithology to survey the peregrine throughout the British Isles. Derek Ratcliffe was approached to head the survey team and to coordinate the data.

As information began to come in from the regional network of observers, it quickly became apparent that the peregrine was in deep trouble. In his book, Ratcliffe wrote:

The once flourishing population of southern England had all but disappeared and that of Wales was greatly reduced. In the north of England, numbers were well down on their normal level, and in coastal southern Scotland few falcons remained. In the Scottish Highlands, too, there were local signs of decrease. Breeding success was low amongst the remaining peregrines and numerous territories were occupied by birds which failed to rear young, or even to lay eggs. Many nesting places were occupied by single peregrines and over the country as a whole a large number of nesting places once held regularly by peregrines seemed to be totally deserted, in many cases for the first time ever known.

The picture emerging was thus in marked contrast to that presented by the pigeon fanciers, and by the end of the first year of the enquiry, the anxiety among nature conservationists was no longer about the homing pigeon issue, but about the apparently headlong decline of the peregrine population and its causes.

The peregrine's problems were not confined to Britain: reports from Europe and North America were horribly similar – eggs failing to hatch, eggs being eaten by the falcon, the disappearance of breeding adults. In *The Falcons of the World* (1982), Professor Tom Cade wrote:

By 1964, in the United States, not a single breeding pair or even a lone adult could be located east of the Mississippi River in a region where nearly 200 eyries had been occupied through the 1940s. Peregrines were also greatly reduced in the western states, particularly in California where probably not more than 10–20 per cent of the state's 100 known eyries still held peregrines, in Utah where all but two or three of the state's 40 known eyries had been abandoned, and similarly in other western states such as Oregon, Washington, Idaho and Montana. Less severe reductions had also occurred in Arizona, New Mexico and Colorado. The falcons were also gone from virtually all of their known eyries in Canada south of the boreal forest.

With a similar picture emerging across most of Europe, an explanation was needed urgently. It was Derek Ratcliffe who discovered the first clue which enabled scientists to identify the problem. He sent failed eggs for analysis. They were found to contain high levels of chlorinated hydrocarbons. This immediately rang alarm bells linking the agricultural use of DDT and dieldrin to the peregrine, who, from her position at the top of the food chain, feeding on granivorous birds, would be highly sensitive to any damage resulting from the use of the organochlorides to dress

crops. Moreover, as further studies were carried out on the remaining population of the species in the British Isles, it became clear that the areas of Britain which had suffered the most severe decline matched the pattern of agricultural concentration. However, scientists still could not pinpoint exactly how the pesticide was causing such widespread failure at nest sites. In 1965 an International Peregrine Conference was held at the University of Wisconsin, in Madison, bringing together those working with peregrines in the field, but still no definitive answer could be found.

Again, it was Ratcliffe who answered the question. He was puzzled by the high incidence of broken eggs which he had found at eyries. He began to suspect that the eggs were thin-shelled, and he prevailed upon those in charge of historic collections to measure the thickness of the peregrine eggshells in their charge, so that a comparison could be drawn with the shells of recently failed eggs. The latter were found to be an average of 20 per cent thinner. Armed with this information, scientists discovered that DDE – a breakdown product of DDT – acted on the enzyme systems within the falcon which dictated the movement of calcium and carbonate – the main constituents of the eggshell. The mystery was solved – but the problem remained.

At the time in her history when the peregrine most needed man's help, it was forthcoming. It was not only the peregrine who was affected by DDT: other raptor species were found to be afflicted through the same route, the results again being manifested in thin-shelled eggs. Conservationists and naturalists began to call for a ban on the use of heptachlorides. Britain spearheaded the campaign, initially with voluntary restrictions on the use of DDT and dieldrin. The results were encouraging: by 1971 peregrines were increasing in number in the UK, and by 1981 they had reached the highest breeding densities known in the twentieth century. In the United States, legal hearings to ban DDT brought the peregrine to the centre of public attention, and her plight touched the hearts of the American people. But would banning the use of the pesticides provide sufficient impetus for the species to regain its stronghold over the now barren territories in eastern North America? Scientists and naturalists feared not.

The extinction of the peregrine in the eastern United States and south-eastern Canada called for desperate measures to be taken. Given the extent of the damage, complete natural repopulation seemed highly unlikely. Thus a man-made disaster needed a man-made remedy. It was Tom Cade who decided on a bold course of action to help the peregrine recover her lost territories. Cultivating the seeds of an idea sown at the Madison conference, in 1970 he founded the Peregrine Fund under the auspices of Cornell University in Ithaca. The aim of the Fund was to breed falcons in captivity and to release them into the worst-affected areas, in an attempt to get them to repopulate these territories. The idea received much criticism, but its supporters were adamant that it was the only realistic hope for the peregrine to recover her previous strongholds.

At the time, breeding of birds of prey in captivity was in its infancy. However, using knowledge acquired through early experiences with peregrines, breeding barns of suitable dimensions were built in Ithaca and these were optimistically stocked with peregrines drawn from a variety of backgrounds. Falconers donated birds, wild chicks were taken from the nest, and, in 1972, four pairs of falcons were brought in from the wild at fourteen to twenty days of age. These were hand-raised in groups, handled minimally after fledging, and paired up in their immature year. The first successes occurred in 1973. The young produced were retained in captivity, together with four birds from the wild, which were placed with adults and allowed to fledge from the ledges from which, it was hoped, they themselves would later breed.

When the captive breeding stock was deemed to be of sufficient strength in terms of numbers and viability, the first chicks were 'hacked back' (released under

*Chicks being raised by the Peregrine Fund.*

controlled conditions) to the wild at selected sites. The chicks were placed on ledges in the wild, at a stage when they could feed themselves, but before they could fly. Every day, food was supplied to the site by means of a long tube or similar device which enabled the hack-site attendants to provide food without being seen themselves. In this way the eyasses fledged just like wild chicks, and began to make early sorties from the ledge. Food continued to be provided until the eyasses were killing for themselves regularly and were no longer in need of supplementary supplies.

As the released eyasses began to spread out across their new territories, some remarkable stories filtered back to the Peregrine Fund about the captive-bred youngsters. One such story concerned a young falcon which was released in 1977 from Carol Island – a wildlife refuge maintained by the US Army and located at the mouth of Gunpowder River in north-east Maryland. From the point of her release,

the falcon made her way into the city of Baltimore and took up residence on the thirty-seven-storey USF&G Insurance building. Here she became the object of much attention, as some 9,000 employees observed her movements closely, and she was named 'Scarlett', after the heroine of *Gone with the Wind*.

The Peregrine Fund assisted in the creation of a nest scrape for Scarlett, to encourage her to stay. Over the next two years, she seemed content to remain based on 'the highest cliff in Baltimore' as she hunted the large urban pigeon population which colonised the city. In 1979, two tiercels were released as potential mates for her. One flew off, while the other failed to appeal and was subsequently recaptured. Meanwhile, Scarlett laid infertile eggs, which were replaced by chicks born in the Peregrine Fund laboratories in Ithaca.

Over subsequent years, Scarlett failed to mate successfully with three further tiercels which were released for her. She produced infertile eggs every year, which were again replaced by captive-bred chicks which she reared and which fledged successfully. Then, after her many observers and supporters had virtually given up hope of her ever raising her own chicks, a new tiercel was seen in the city in July 1983. This individual had no leg-bands and was therefore clearly a wild bird, whose origins could only be guessed at. They paired up, and in the following year history was made as the first chicks known to have hatched outside of captivity in an eastern North American city since the 1950s fledged successfully and left the city to seek out their own territories.

This was the hallmark of success that the Peregrine Fund had been waiting for. Reintroduced peregrines had produced chicks in the wild, and the project had proved its worth – vindicating its founders in the process, for they had proved conclusively that aviculture could indeed be used as a tool to reintroduce a threatened avian species. The continuing success of the project has resulted in the seed stock of wild-breeding peregrines doubling their numbers every year or two in eastern North

America. Using the knowledge gained with the peregrine, the Fund has now been able to turn its attention to other endangered raptor species, founding the World Center for Birds of Prey, in Boise, Idaho.

Further afield, as the full implications of the effects of DDT were realised, the tide of public opinion turned back in favour of the peregrine, and efforts were also made to help her in other countries. Self-funding protection organisations specialising in birds, such as the Royal Society for the Protection of Birds in Britain, started to monitor nest sites to protect them from unwelcome and illegal attention. While British peregrines had recolonised on their own, those in other parts of Europe needed specific help. The Arbeitsgemeinschaft Wanderfalkenschutz was founded in West Germany in 1965, with the aim of identifying and removing the causes of the peregrine's decline there. Their findings prompted the German falconers' club, the Deutscher Falkenorden, to start a reintroduction programme in 1977. Artificial tree nests were built to emulate the natural nest sites on the north-German plain, where peregrines nested exclusively in mature pine or beech stands. Captive-bred peregrines were also hacked from high buildings to encourage the eyasses to select similar secure nest sites when they reached maturity.

Falcons released in West Germany moved into East Germany at a time when the peregrine was bordering on extinction there. The nest sites of the few remaining pairs were vigorously protected, and artificial nest platforms were constructed, old nests were repaired, and protection was provided from the weather. Across Europe, help of one sort or another was given to the peregrine wherever it was deemed prudent, and gradually her numbers recovered.

In 1985, a further conference on the peregrine was called in Sacramento, California. Over five hundred biologists and peregrine specialists met to assess the worldwide status of the peregrine and to celebrate her survival. Papers from as far afield as Australia, the Far East and South America, and from remote outposts such

as Fiji, were presented to the largest-ever gathering of peregrine aficionados. The delegates learned that the peregrine was indeed a survivor, for a mere twenty years on from the Madison conference – which many had feared might be a requiem for this remarkable species – she had overcome ecological disaster and provided living proof that man, through active conservation measures, is capable of righting a wrong and reversing the progress of a species caught in a seemingly headlong slide towards extinction.

# THE CAPTIVE BREEDING OF PEREGRINES

*Clearly the range of relationships and adjustments that are possible for peregrine falcons in a world populated by sympathetic human beings is far greater than anyone could have predicted …*

Professor Tom J. Cade, *Peregrine Falcon Populations, Their Management and Recovery*

It was falconers who pioneered the breeding of raptors in captivity. Up until the mid 1960s, it was considered so difficult to persuade a pair of birds of prey to reproduce in an artificial environment that few people bothered to try. Zoos displayed single raptors, merely as an example of the species, and falconers acquired their hawks from the wild, in accordance with the laws of their lands. It was only when wild hawks became threatened that falconers decided that they must turn their hand to aviculture. Their desire to know intimately the species which they so loved and admired coupled easily with a simple wish to have a guaranteed annual supply of eyasses. They pursued their aim for a self-sustaining captive population with an obsessive effort which was, in the case of the peregrine, to pay rich dividends in her future.

Historically, there appear to have been few attempts to breed peregrines in captivity. Tom Cade of the Peregrine Fund has researched the subject and has found two cases worthy of mention. A. G. Johnston from Dumfries, Scotland, mentioned in a communication to *The Naturalist* (1853) that a pair of peregrines, having been kept in confinement for some years, laid two fertile eggs in 1852. Apparently both tiercel and falcon incubated for twelve days, but were put off by human disturbance. The fertility of the eggs was proven by examination.

The other record of interest comes from the United States and concerns a data card found with a clutch of peregrine eggs in the collection of the Western Foundation of Vertebrate Zoology. It reads:

> Collector: C. Nicols for C. Littlejohn, Redwood City
> Locality: Sea Shore, West of Pescadero, California
> Date: 21 March 1897.

In 1896 four young were taken from the same nest and kept in confinement, when grown two were given liberty and one of the remaining two laid at least two eggs in the house in which they were confined. I did not learn the exact date but it was about the same time the old birds deposited this set.

The German falconer, Renz Waller, was the first person on record to succeed in producing captive-bred peregrines. In 1942, in Düsseldorf, during the Second World War, a twelve-year-old falcon captured as an eyass laid fertile eggs. The tiercel was a passager which had been in captivity for many years. He had a wing injury which prevented him from flying. One of the eggs hatched and the chick fledged. The pattern was repeated for several years, each time with a single chick being produced.

Waller's story is remarkable, because the odds against a severely injured tiercel mating with a falcon amid the hail of Allied bombs which were raining down on Düsseldorf in 1942 must be considered to have been extraordinarily high. It inspired many other attempts to breed peregrines in captivity in Europe and America, but none was destined to succeed until significant results had been achieved in the 1960s with a captive population of American kestrels. Through building up breeding colonies of the kestrel, early pioneers in raptor propagation learned much about the requirements necessary for success with other species of falcon. The production of second-generation stock from the kestrels proved conclusively that it was possible to achieve production of eyass raptors from a captive population on a large scale. These results further inspired those who were, by this time, certain that the regular production of peregrine eyasses from pairs held in captivity was a realistic goal.

When ecological disaster hit the peregrine on a global basis, the desire to breed peregrines in captivity became invested with an urgency not previously anticipated. Pinning his hopes purely on the results achieved by Renz Waller and those achieved with the American kestrel, Tom Cade put forward his bold plan to breed and release

*The stages in the peregrine's development from egg to adult.*

peregrines, to a splinter group of falconers and biologists which assembled after the 1965 International Peregrine Conference in Madison. They forged plans to obtain wild stock for breeding, and in 1966 founded the Raptor Research Foundation Inc. to act as a clearing-house for the dissemination of information between breeders. To this end, the Foundation published *Raptor Research News*.

In 1966, the Raptor Research Foundation was permitted by the provincial authorities of British Columbia to take several pairs of young Peale's falcons. These were subsequently distributed to private breeding projects. When these pairs reached maturity, the results were mixed, but nevertheless encouraging, for one chick was raised to maturity in 1968, and in 1971 Dr Heinz Meng succeeded in hatching a single chick by artificial incubation. This tiercel was called 'Prince Philip', and he was deliberately imprinted on human beings, so that he could subsequently be used as a semen donor for artificial insemination. Dr Meng capitalised on his success by producing a further seven chicks from the same adult pair the following spring. He loaned the breeding pair to Tom Cade's recently formed Peregrine Fund, together with 'Prince Philip', and thereby provided a significant boost to the project, for they were able to rear twenty-three eyasses from the original pair during the next five years, and 'Prince Philip' fathered many chicks in his capacity as a voluntary semen-donor. In 1973, the Peregrine Fund produced twenty eyasses from three pairs, at the same time as other projects in Colorado and Alberta also began to bear fruit. Projects in Europe announced success from 1971 onwards, with notable projects including in particular the one run under the auspices of Dr Christian Saar in Germany.

Scientists have always needed terrific reserves of determination and conviction to persuade others of the merit of their ideas, and Tom Cade was no exception. As the breeding programme became increasingly successful, his obsession with the repopulation of the peregrine in the wild enabled him to overcome his opponents and detractors and to bring about the breed-and-release scheme which led to the rebirth of the peregrine in America.

The first releases of captive-bred peregrines took place in 1974, when two eyasses were hacked back to the wild. These were the first of thousands. After the successful natural mating of 'Scarlett' and the wild tiercel in Baltimore in 1984, the scientists finally knew that their goal was achievable. In 1985, five out of nine known pairs formed from released birds raised chicks in the wild, and the recovery of the peregrine in the eastern states of North America was well under way.

The experience gained by the Peregrine Fund was publicised to falconers and aviculturists around the world. Those who had previously merely put a falcon and a tiercel, selected at random, together in an aviary and hoped for a result had a lot to learn, as it became apparent that any results thus achieved had involved a large measure of luck.

The original stock at the Peregrine Fund's headquarters at Cornell University could be placed in three main categories according to their backgrounds. There were passage hawks – trapped from the wild during their immature year and acquired from falconers. There were eyasses taken from wild nests at under four weeks of age and crèche-reared by hand in groups, then kept in the manner for falconry, before being paired up. Finally, there were four pairs brought in from the wild at fourteen to twenty days old and handled minimally before being paired up.

As soon as the breeding programme was successful, the Fund could create fourth and fifth categories. One group was made up from captive-bred eyasses which fledged under pairs and were subsequently put into large communal chambers, in the hope that natural pair bonds could be identified. Those in the fifth category were reared in the same way, but were handled during their immature year as they would be for falconry, being tethered, hooded and fed in close proximity as pairs, but without being flown.

After the intensive study of the success rate of all these pairs, in relation to their origins, the Fund became able to predict the likely future breeding success of pairs formed in these ways, and to adjust their future pairings accordingly. Two methods of forming pairs were clearly less successful – the trapped passager stock produced poor results, and the captive-bred peregrines which were not handled using falconry techniques were extremely nervous.

As clear-cut guidelines for forming pairs became apparent, so the breeding programme progressed in leaps and bounds. Many of the gaps in the knowledge of the peregrine's natural biology were filled in, as details of courtship, copulation, egg-laying and incubation were minutely observed by the technicians at Cornell. This information was disseminated across the world, and was not only helpful to the growing number of 'backyard' breeders, but also of great interest to ornithologists and naturalists worldwide.

As the project progressed, the scientists at the Peregrine Fund became more adventurous, making inroads into the psychology of falcons 'imprinted' on humans and their degree of usefulness within a breeding programme.

Any raptor which is reared in isolation by hand will imprint on people. Those which are merely fed by hand will scream at people for food and attention, in the same way that they would demand food from a natural parent. In falconry, such birds are painful to train as they have atrocious manners because, as a result of early familiarity, the barrier of deference which exists between a falconer and a parent-reared falcon has never been formed. In breeding terms, too, 'food imprints', as they are called, are useless, for they have never learned to socialise with other falcons. They therefore not only do not recognise other falcons as potential mates, but they can also be extremely aggressive towards their own kind.

Tom Cade recognised quite early in the programme that falcons which were 'correctly' imprinted could, however, be used with a high degree of success for artificial insemination. 'Prince Philip', the imprinted tiercel loaned by Dr Meng, had proved this. Efforts were therefore made to imprint a selected number of tiercels to act as semen-donors, and similarly to imprint falcons which would accept insemination. To produce such birds, the eyasses were reared by hand, but the technicians spent time socialising with them as well as providing food. Borrowing techniques from falconers, the stock thus reared became fixated on their human companions both physically and psychologically. In the breeding season, the conditioning of these imprints included the technician going through many of the courtship procedures which occur between a natural pair – complete with vocalisations and food-passes.

The use of artificial insemination in the captive breeding of peregrines opened up new horizons in terms of productivity. By maintaining imprinted falcons on their own in suitable chambers, the technicians could condition them to produce between

*TV monitoring at the Peregrine Fund enables the technicians to observe the breeding pairs without disturbing them.*

A Roll X incubator containing peregrine eggs.
Eggs that are incubated naturally for the first
week to ten days before being transferred to an
incubator are more likely to hatch.

eight and fourteen eggs, each fertilised artificially during the interval between one egg and the next. Imprinted falcons also make good foster-parents, so, following the pattern devised for naturally laying pairs, the eggs could be removed for hatching in an incubator, to stimulate further egg-production, with the resulting eyasses being returned at a later stage to undergo the natural rearing process necessary to render them fit candidates for release.

As the Peregrine Fund energetically explored the new avenues which opened to them each time they mastered a fresh technique in the field of captive breeding, so the backyard breeders too began to produce sufficient numbers of eyasses to supply high-quality peregrines commercially to falconers on an annual basis. Artificial insemination enabled breeders to hybridise between different species of the genus *Falco*. In this manner they produced hybrid prairie/peregrine falcons, gyrfalcon/

*Eyasses branching out from a hack-box.*

peregrine falcons, saker/peregrine falcons and many others – even crossing peregrines with the smaller falcons such as merlins. The falcons thus produced displayed 'hybrid vigour' and often embodied the best traits from both parent species.

In seeking to produce a hybrid falcon ideally suited in terms of size and performance to taking a particular type of quarry, the backyard breeder has begun to follow in the footsteps of breeders of domestic animals, who have long been practising selective breeding in order to produce the finest horses, dogs, cattle and so on – in some cases for many centuries. While it seems doubtful that man can actually improve on the perfection of physical form which nature has already given the pure peregrine, within the species itself breeders are now beginning to identify strains which lend themselves more readily than others to the tasks set for them by falconers. Today, therefore, if a falconer requires a peregrine, he will attempt – if he is knowledgeable about such matters – to acquire one from a pair whose progeny have proven equal to the intended task.

As the science of raptor propagation progresses, the pedigree of captive-bred peregrines will undoubtedly come under increased scrutiny. Meanwhile, both peregrines and man have undeniably benefited from the passionate endeavours of the handful of men who, with such unwavering determination, pioneered the early efforts to turn the captive-bred peregrine from a dream into a reality.

*The Peregrine Fund, Boise, Idaho.*

# THE HISTORY OF THE PEREGRINE IN FALCONRY

*The peregrine riding the wind, looking down on the world from above – or gentled and caressed on the falconer's wrist – with its finely worked hood, its crest of feathers, its distinctive plumage, with those cruel weapons which destroy inferiors at a blow, is almost a physiological extension of its master: an image, conscious or unconscious, of the knight helmeted and armed in the panoply of the late-medieval passage of arms.*

Pero Lopez de Ayala, Chancellor of Castile, *Libro de la Caza de las Aves*

The vast majority of early references to the peregrine relate to the ancient art of falconry – in the words of John Cummins 'an intrinsically superior activity, an appropriate pastime for the aristocracy' – wherein the peregrine was widely considered to be 'the noblest and best of the birds of prey, the lord and prince of hunting-birds' (Pero Lopez de Ayala). While describing the peregrine within falconry, early writers also documented something of her natural history – if only to describe the habitats in which strains to be taken for falconry could be found. Many of the references are brief, listing the peregrine alongside other hunting-birds used within the sport, such as the goshawk, the gyrfalcon, the lanner falcon and the sparrowhawk. However, such were the virtues of the 'falcon gentle' that they are considered worthy of frequent mention from which it is possible to glean from this a little of the peregrine's natural and unnatural history.

The earliest remains of the genus *Falco* have been dated to the Miocene period, some twenty-five million years ago. Actual peregrine remains have been identified from fossils or subfossils, found in British cave earths in Somerset and Derbyshire. These have been dated no more specifically than the late glacial period. The earliest records of falconry, however, can be dated back to around 1400 BC. A bas-relief from the period, found in the ancient ruins of Khorsabad, Iran, depicts a falconer with a hunting-bird on his fist, standing beneath another bird stooping at quarry. Like so many of the early animal and bird depictions, it is not possible to identify which species of hawk the carving is intended to portray.

One of the earliest depictions with which the peregrine can be specifically identified is a hieroglyph dating from the ancient Egyptian civilisation: a manuscript belonging to the lady Meri features a drawing showing a falcon with a black cap and moustache, sitting on a perch, with what might be a lure at her feet. The Egyptians worshipped Horus, the sun-god, and built many effigies of him, including the gigantic statue which guards the entrance to his temple at Edfu, built by Ptolemy in

50 BC. Many of the images of Horus appear to be based on the peregrine – some feature the familiar black cap and moustache – while others appear to resemble more closely a kestrel or a lanner.

While these drawings and carvings are some evidence of the peregrine's early presence in North Africa, actual documented records of her do not appear in the West until the eighth century.

It is difficult to estimate when the first peregrine was trained in the West. Peregrine bones have been found in excavations of Viking dwellings dating from the ninth and tenth centuries, and there are records showing that from 1173 Henry II of England sent annually for young peregrines from the sea cliffs of Pembrokeshire in Wales. It is, however, certain that the peregrine was among the most popular of hunting-birds in the Middle Ages. During this golden era of falconry, the sport was widespread and was enjoyed by commoners and noblemen alike.

In 1486, Dame Julians Barnes's *Boke of St Albans* laid out a table indicating 'the naamys of all maner of hawkys and to whom they belong'. There followed a long list of recognised ranks and statuses, linked with species and types of hawk. In the list there appears an entry which reads 'This is a *falken peregryne*. And that is for an earl.' The list further refers to the '*falken* of the rock' for a duke and the '*fawken gentill*' for a prince, both of these being alternative names for the peregrine. Many historians have interpreted this list as a code of ownership for the various species of hawk, but this has perpetually puzzled falconers, who know that many of the species mentioned are useless within falconry. In his book *The Hound and the Hawk: The Art of Medieval Hunting*, John Cummins puts forward the theory that the list relates to imagery – the behaviour of each species being related to the standing of various ranks at court – rather than reflecting a code of ownership to which the ranks were expected to adhere.

The first in-depth study of the peregrine appears in a remarkable and lengthy volume by the Holy Roman Emperor Frederick II of Hohenstaufen, entitled *De Arte*

*The traditional method of carrying falcons in the field on a cadge has remained unchanged since the Middle Ages.*

*Venandi cum Avibus* (1248–50). Within this work, as well as revealing his passion for falconry, Frederick also displays a detailed and impressive knowledge of the species employed within the sport at the time. He devotes a large section to the peregrine, which, for the first time in her written history, is described in terms of different types and colour phases. In addition, Frederick discourses with authority on the potential ability of these different types of peregrine, and writes in great detail about the various methods of training, to suit the purposes for which they were required:

Peregrine falcons exhibit a variety of types. Many of them have well-shaped limbs adorned with fine plumage, while the limbs of others, though shapely, have unattractive feathers. Some peregrines whose plumage is beautiful have ill-formed members; while a fourth class may possess both abnormally shaped bodies and defective plumage. These four classes are not equally commendable.

Subsequent sections deal with training, weather, quarry and 'eight modes of behaviour of falcons sent up to wait on'. Much of the advice given still holds good today, and it is awe-inspiring to consider how little that the nature of the task facing contemporary falconers and the methods they must employ to train the peregrine have changed since the thirteenth century.

From the early thirteenth century, for over four hundred years, falconry flourished both in Europe and the East. Marco Polo reported that in the latter half of the thirteenth century the Chinese emperors of the Mongol dynasty, successors of Gengis Khan, had 'the best falcons and the best dogs in the world'. Every year around March Kublai Khan, grandson of Gengis, went hunting in Manchuria, accompanied by ten thousand falconers and an equal number of guards. The expedition was organised with great precision and surrounded by incredible luxury. The emperor's wives also hunted and had their own falcons.

Falconry flourished in Turkey – a fourteenth-century book on the subject, written by the prince of Meteshe, states that Murad II kept seven thousand falconers. Similarly, the sport was popular in the court of the khans of Persia during the Mogul dynasty. Later, the Shahs continued to display passionate enthusiasm for falconry – it is documented by Chardin that in the seventeenth century, eight hundred birds of prey, each with its own officer, were always kept at the king's venery.

The crusaders learned much of interest from oriental falconers, exchanging ideas, methods and hawks. In the Far East, falconry had been recorded in Japan since AD 239. Many elaborate routines, centred on ceremony and formality, evolved around the sport, because hunting birds belonged to the nobility. Although the peregrine was not as popular in Japan as the goshawk, the Chinese and Japanese Encyclopedia of 1714 refers to one of the names for the peregrine as *nade-taka*, or in Chinese *fou-yng*, meaning 'falcon that one caresses'.

In 1243, Henry III gave falcons from the Isle of Lundy in the Bristol Channel, off the coast of England, 'to his beloved cleric Ade de Eston'. Peregrines from Lundy have been favoured by falconers for many centuries, for their high-flying qualities, and in the early twentieth century they were still in vogue. Roger Upton, in his book *A Bird in the Hand* (1980), describes the famous hawks of the notable falconer Captain Gilbert Blaine in 1900:

Among his hawks was 'Lundy', the first of many peregrines that he obtained from the island over the next twenty-five years. Indeed perhaps this tiercel was the very reason for Blaine's determination to get Lundy peregrines, for in his first season this lovely little tiercel killed the top score of fifty-six partridges, bettering by fourteen the score of the experienced old tiercel 'Ready'.

It was easy for falconers to identify strains that flew well and to continue to take them from the same nest site year after year, for some eyries were occupied for four successive centuries. On estates where peregrines bred, young falcons were frequently demanded as part of the rent.

The hawk of rulers and the monarchy, the peregrine appeared in the mews of all the European rulers who practised the sport of falconry in its heyday. Alexander III, King of Scots from 1249 to 1286, is recorded as keeping falcons in Stirlingshire, and Mary, Queen of Scots from 1542 to 1567, had falcons taken from the wild in the same area. From 1488 to 1508, the names of forty-eight royal falconers appeared in the accounts of the Lord High Treasurer to James IV of Scotland. The desire to possess the best hawks brought out the worst in the nobility, who, during various periods in history, punished the theft of a hawk by death or excommunication and decreed that ownership of the 'noble' hawks should be limited to the aristocracy. Derek Ratcliffe says of the peregrine that:

During the Middle Ages, the peregrine was the sovereign's or the nobleman's hawk, and protected by harsh penalties. For centuries it was mentioned only as an object of esteem, ably assisting man in the chase and deserving only of the greatest admiration and affection for its prized qualities of hunting skill, killing power and courage.

After the Middle Ages, the sport of falconry not only remained popular but enjoyed a new lease of life as traders and adventurers travelled the globe, exchanging ideas about hawks, equipment, training-techniques and related activities. A seventeenth-century Japanese manuscript describes and illustrates the capture of peregrines using a bow net trap. This design was favoured by the Dutch and is likely to have been passed on by the Dutch traders who travelled to Japan in the sixteenth and seventeenth centuries.

*Most peregrines kept for falconry like to bathe vigorously every day, dipping their wings to scoop water over their backs.*

*The hood covers the falcon's eyes so that she may
be carried quietly in the field.*



During this period, falconry was in decline in Europe as a result of the endless wars following the Reformation. The Puritan regime in Britain was less tolerant of pleasurable sports and pastimes, and the invention of the shotgun meant that game could now be taken by an easier method. European birds of prey, including the peregrine, began to be looked upon in some quarters as being in competition with hunters for game.

In the early 1770s, the first hawking club – the Falconers' Club – was founded in Great Britain. By 1783 the club owned 32 peregrines, 13 goshawks and 7 gyr-falcons. Fifty or so club members met each April to fly their hawks at kite and rooks. The club remained in existence until 1838. In 1839 the Loo Hawking Club was established, taking its name from the royal Dutch estate over which the club was to fly hawks, and a number of former members of the Falconers' Club transferred their allegiance to the new club, which enjoyed royal patronage. The subscription was 100 florins per annum, and the principal quarry was heron. Adrian Mollen, a member of the celebrated Mollen family, who were famous for their ability to trap and supply fine peregrines from the plains of Valkenswaard to falconers across Europe, was among the professional falconers whom the club employed.

The authors of the celebrated *Traité de Fauconnerie* (1844) – Hermann Schlegel, an eminent zoologist, and A. H. Verster de Wulverhorst, the author of two earlier Dutch titles on hunting – were both members of the Loo Hawking Club, and within their book they had much to say on the subject of the peregrine, whom they preferred to refer to simply as 'the falcon' or 'the common falcon':

Falconers do not usually designate this species by a particular name because it is the most widespread species of the genus and the one most commonly used for hawking in Europe. In using the name 'common' or 'ordinary' falcon, we have followed the example of Dutch, English, German and Danish falconers, who, when comparing this falcon to the other species of the genus, customarily confer on it an analogous denomination. As we explained . . . it was a mistake for the modern naturalists to apply to this species in general the epithet 'peregrine', which was invented by falconers simply to designate falcons from certain localities taken at a certain period of their passage.

The authors then proceed to describe the peregrine in her European form and to compare forms found in other parts of the world, by size and colour. The name 'Barbary falcon' is specifically mentioned, and they also refer to travellers who 'reported from Java and the Molucca Islands common falcons remarkable at maturity for the dark colour of the ear region'. Thus it can be seen that the species of *Falco peregrinus* was, by this time, beginning to be recognised in some of the subspecies by which we know her today.

During the siege of Paris by the Prussians in 1870–1, peregrines were reputedly used to catch message-carrying pigeons *en route* to the exiled French government in Tours. This was the first of several instances in history when the peregrine was used for this purpose.

In the late nineteenth and early twentieth centuries, falconry clubs continued to be formed in Europe. When the Loo Hawking Club died out in 1853, the next club to be formed was the Old Hawking Club in Britain in 1864. The primary aim of this club was to fly the peregrine at rooks on the Wiltshire Downs, with seasons at heron and grouse following. The club succeeded magnificently in this aim and enjoyed a high standard of sport for sixty-two years. Following its demise, in 1927 it was succeeded by the British Falconers Club, which is still in existence today.

The Deutscher Falkenorden was formed in the same decade, 1923. Ten years

*A falcon kills a grouse by severing its neck with the 'tooth' in her beak.*

later the club received the patronage of Field Marshal Hermann Goering, and the following year a state falconry centre – Der Reichsfalkenhof – was built in Riddagshausen in Brunswick, housing many peregrines together with other hunting-birds.

Cooperation between clubs became common in the 1960s, and jointly they helped to orchestrate protection for their native raptors. Under various laws concerned with wild birds, the peregrine gradually achieved protection across Europe, and her decline – brought about virtually exclusively by human persecution – was halted. It is clear that, whether prized, damned, persecuted or protected, the peregrine has survived the course of history in the wild and in the hands of man. Always an inspirer of strong passions, she will no doubt pass through many similar episodes in her future.

*Peregrines 'weathering' on the lawns of Braco Castle, Perthshire. The weathering lawn is traditionally the place where falconers' hawks are put to bathe and enjoy the elements.*

# THE PEREGRINE IN CONTEMPORARY FALCONRY

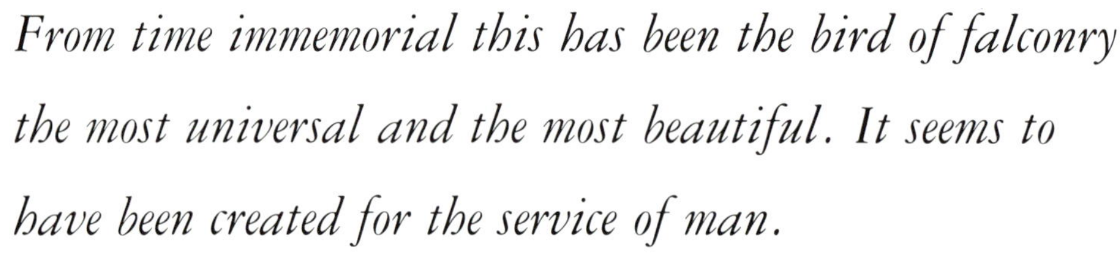

*In the stoop the peregrine will form the characteristic anchor shape before closing her wings tight to her body for acceleration.*

For centuries, the peregrine has been held in the highest esteem by falconers. Although she is neither as large nor as fast as the largest of all the falcons, the celebrated gyrfalcon, the supreme sporting balance between the peregrine and her traditional quarries has earned her a special niche in the falconer's affections and has guaranteed her a place in his mews.

For the vast majority of falconers across the world, the peregrine epitomises everything they want in a hunting-bird. She can be versatile or highly specialised, according to the nature of the opportunities with which she is presented. As a hunter of other birds, she can be trained to work in two distinct ways – 'out of hood', flying straight off the fist in direct pursuit of quarry, or 'waiting on', when she is taught to leave the fist and climb steadily to a good height, before game is flushed beneath her. Both styles of flying can incorporate her stoop – a word which seems inadequate to describe the sight of a falcon hurtling earthwards with her wings pulled close to her body, seemingly risking an untimely death should she fail to pull up in time. Within these two styles of flight, falconers around the world have trained the peregrine to hunt the fastest of their native birdlife, and have forged with her an enduring partnership.

Over the centuries, the sport of falconry has developed its own language, in which some of the terms used enable a falconer to identify the origins of a hawk and thereby to predict the style of training necessary to achieve the best performance from it. A hawk taken from the nest is known as an 'eyass'. A hawk trapped during her first year is called a 'passager' or 'passage hawk', and one trapped in its adult plumage is known as a 'haggard'. In terms of the peregrine – as indeed with the majority of the species of birds of prey used for falconry – the passage hawks have always been the most favoured. The reason for this is straightforward – the passager has learned to hunt in the wild, and is therefore much more competent than an eyass, which knows nothing of hunting, has had no time to practise her flying skills, and has no muscle or fitness.

While the haggard shares with the passager all of those qualities lacking in the eyass, she is much more difficult to train and more easily lost, having matured in the wild and developed her own preferences in terms of quarry. Thus she is much less tractable

*Adult (left) and immature tiercel, showing the difference in tail length, which shortens after the first moult.*

than a passager and frequently will not countenance the type of quarry to which the falconer would like her to devote her considerable skills.

In modern times, some countries which protect their native birds of prey will no longer allow hawks to be taken from the wild. In Britain, for example, the only peregrines which can be used within the sport of falconry are those bred in captivity. Thus falconers in Britain have to be adept at training eyasses, guiding them with patience through their formative training in the early months until they are using their inherent aerial ability to the full. In some other parts of the world, including parts of Africa and the Middle East, passage peregrines can be trapped. Although these will be quicker to achieve initial success in the field, the end result will be no different from the performance of an eyass with a few months of experience behind her.

By watching wild peregrines hunt, early falconers were able to identify and select the two styles of flight previously mentioned – direct pursuit and waiting on – as being the most suitable for their purposes. Before embarking on the training of a peregrine, the falconer will predetermine the quarry which he wishes her to catch and therefore the style of flight which she must be taught to adopt. Direct pursuit is the easier of the two to teach, as it is a peregrine's natural instinct to pursue a suitable quarry species in flight. Waiting on is much harder to teach, as the peregrine is expected to make a huge effort to gain height before she is even shown the quarry.

The difference in training to achieve waiting on or direct pursuit does not occur immediately. The early stages which include manning, when the falconer carries the peregrine, hooded and unhooded, on his gloved fist, feeding on the fist, and calling off to the lure – a pair of bird wings attached to a length of line, designed to resemble the intended prey – are exactly the same in both cases. The difference comes when the peregrine is ready to fly free, having been responding reliably to the lure over a distance of fifty to a hundred yards while still on the creance – the light line used to secure her during training. At this point, a hawk intended for direct pursuit will be flown free to the lure, which will always be kept visible. Either she will be called to it over extended distances, or she will be encouraged to 'stoop to the lure'.

Stooping to the lure teaches persistence and is an excellent way of building up fitness. The falconer swings the lure in clockwise circles at his side and offers it to the falcon as she approaches, drawing her past him, then removing the lure from her flight path at the very second she is expecting to catch it in mid-air. After a number of passes to the lure, the peregrine is rewarded by being allowed to catch it in her feet. The number of stoops or the distance over which she is called is increased daily, and in this way the peregrine is fit enough to 'enter' at quarry within a week or two of flying free.

A peregrine intended for waiting-on flights is encouraged to wait on by the

*Immature falcon waiting on.*

falconer hiding the lure once she is on the wing, causing the hawk to circle him, awaiting the reappearance of the lure. Flights at game from the waiting-on position are most usually successful if the game is flushed downwind, thereby allowing the peregrine to take full advantage of the wind to augment the speed of her stoop. Thus game hawks, as those which are successfully trained to wait on are called, must be encouraged to stoop downwind right from the earliest days of flying free. The falconer must time his throwing out of the lure to coincide with his hawk being upwind, and preferably as high above him as possible, so that she learns that she will be rewarded from this position.

Unfortunately, although a peregrine may be easily taught to circle the falconer, and to stoop downwind to the lure, she may never learn to gain sufficient height to enable her to be consistently successful at game. The 'pitch' or height at which a falcon waits on largely depends on the hawk herself. Many consider that game hawks are born, not made, and a natural high-flyer is a gem to be treasured, for she will spare the falconer the drudgery of showing her endless game which she fails to catch through being at too low a pitch. While some peregrines will learn to mount through missing game from a lower pitch, many more will simply fail to make the grade and will have to be tried at less demanding quarry.

The pitch of an established game hawk can vary from 150 to over 1,000 feet. If the sight of her 'towering in her pride of place' high above the falconer is a magnificent spectacle, the stoop which follows waiting on from a high pitch can literally take the breath away. Although from such a commanding position she is likely swiftly to overhaul the game, the kill is by no means certain, as the peregrine's traditional quarries have spent many centuries evading her stoop, and frequently demonstrate such fine tactics of evasion that the spectator can do little more than marvel at the natural sporting spectacle which has unfolded before his eyes.

The quarry at which the peregrine is flown varies from country to country. Generally, those trained to catch any game bird – for example red grouse in the UK; pheasant and partridge throughout Europe; sharp-tailed grouse, sage grouse, duck, partridge and pheasant in the USA; and francolin and teal in Africa – will be trained to wait on. Those trained to catch non-game species – such as rooks and crows, seagulls and, in the Middle East, the *houbara* bustard and the stone curlew – are trained in direct pursuit.

Rook-hawking is a classic example of direct pursuit. The falconer, with his peregrine hooded on his fist, will look for a small group of rooks feeding on the ground in open country upwind of his position. When his proximity causes the rooks to take to the wing, he will remove the hood and the falcon may bob her head, to indicate that she has seen them, before leaving the fist. An experienced rook hawk will not head directly for the rooks but will fly out to one side, climbing as quickly as

possible to gain the advantage of height. The rook will attempt to outfly her into wind. The peregrine will aim to achieve a position both higher and further upwind than the rook. If she achieves this dominant position, she will 'shepherd' the rook, turning it downwind, whereupon, finding itself at the mercy of her stoop, it will flee desperately for ground cover. When the peregrine stoops, the rook will try to evade by jinking. More stoops follow, as the peregrine stands on her tail to regain height and presses home the attack once more. The battle ends when the rook is either struck in the air or achieves the sanctuary of cover on the ground.

The best flights at rook occur when the rook tries to outfly the falcon by 'ringing up', climbing in circles as the falcon rings up beside it. Such flights can reach great heights and cover many miles before the victor is decided. The nature of rook-hawking is such that it does not achieve big bags – if a peregrine manages to catch twenty to thirty in a season, the falconer will be delighted. The highest total of rooks ever to fall to the foot of one falcon in a single year is seventy-two. This record has been held since 1913 by a passage falcon that belonged to the Old Hawking Club, named, aptly, 'Aimwell'.

One of the best-known forms of the waiting-on flight is adopted in the élite sport of grouse-hawking, which takes place in the UK, on the moors of northern England, Wales and Scotland. Red grouse are a worthy adversary for the peregrine, being one of the swiftest game birds, with an impressive repertoire of escape tactics at their disposal – the legacy of the many centuries that their ancestors have spent avoiding the wild peregrines which inhabit the same moorland landscapes.

A peregrine can catch grouse consistently only when she is stooping from a high pitch. To enable her to reach such a pitch, a falconer will seek an area of moorland with updrafts, which will help to give his hawk a degree of natural lift. Pointing-dogs are used to locate the grouse, which cannot be seen when tucked deep into the heather. The positioning of the peregrine – high and upwind of the grouse – and the

*Grouse hawking on the Perthshire moors. The
falcons are carried on a cadge so that the falconer
can select his hawk for each flight.*

timing of the flush are critical, for, if the hawk is out of position, she may stoop but
she is most unlikely to make contact.

The sport begins when a dog is released to range across the moor, quartering into
the wind in search of scent. When it locates a covey of grouse, it goes on point –
stiffening its tail and indicating with its nose the direction in which the grouse are
lying. The falconer unhoods his hawk and she leaves the fist to gain height. If she is
an experienced game hawk, she will waste no time but will begin to mount over the
falconer as rapidly as possible. It is important that she stays overhead during this
period, as her presence will ensure that the grouse stay in the heather, rather than
making good their escape before she is in position.

When she has climbed to reach her pitch, she sets her wings and hangs in the air
above the falconer, who has meanwhile made his way around the point and is waiting

upwind, in a straight line with the grouse and the dog's nose. The grouse now lie between the falconer and the pointer and the stage is set. On command, the dog runs in and flushes the grouse, or the falconer flushes the covey himself, shouting to his hawk, so that she turns over on the wind and stoops. As her feathers tear through the wind, she plummets down, overhauling the grouse until she strikes it down into the heather, or comes up behind it and binds to it, bringing it down to earth in her feet. She is rewarded by being allowed to plume her prize, eating the head and neck, before she is picked up and hooded and the grouse is put in the game-bag.

Grouse-hawking is not an easy sport, for it requires perfect teamwork between the hawk, the falconer and the dog. The grouse are unpredictable and will use any ruse at their disposal to elude the hawk – braking just as the falcon is about to make contact, or dropping like a stone into the heather beneath her stoop, so that she passes

*Immature* Falco peregrinus calidus.

*Female peregrine on a Macqueen's bustard.*

harmlessly overhead. The weather, which over the course of the grouse-hawking season can change from balmy August days to blizzards in the bitterly cold Scottish winter, also plays its part, as in cold weather the grouse become less inclined to lie to the point, sometimes taking to the wing long before the dog has been able to pinpoint their position.

In addition to the challenges posed by the sport itself, there are practical difficulties for the aspirant grouse-hawker to overcome. With the dramatic decline in the numbers of red grouse during the twentieth century, good moors have become increasingly hard to come by and expensive. Guns will pay a lot of money to shoot in a day numbers of grouse which would keep the average falconer going for several seasons. But, despite the many problems which must be overcome in order to pursue this branch of falconry, the sight of a peregrine stooping at a grouse, against a backdrop of dramatic Scottish scenery, is a vision which those fortunate enough to witness will treasure.

In 1968, a falcon called 'Bitch' belonging to Stephen Frank, a master grouse-hawker who has dedicated much of his life to the pursuit of the sport, took 177 grouse and one snipe in the course of a single season. This record is unlikely to be beaten.

In the Middle East, peregrines, which are there called '*behri*' or 'shaheen', are held in the highest esteem. Together with the saker falcon, they have been the mainstay of Arab falconry since pre-Islamic times. The older Arabs prefer the peregrine to the saker, having become familiar with her when sakers were still extremely rare in the Middle East. Arabs generally look for the largest, most powerful peregrines they can find, as the *houbara*, or Macqueen's bustard, is a formidable adversary, more suited to the larger falcons. About 60 per cent of the peregrines in the Middle East are *Falco peregrinus calidus*, and 40 per cent *Falco peregrinus peregrinus*, although most of the subspecies have been tried at one time or another. The local peregrines are highly

favoured in the Gulf states, because of their higher tolerance to humidity. Some Arabs fly *tiba*, or tiercels, mainly at *kairowan* – the stone curlew. (Many Arabs still believe that the smaller tiercels are in fact the females. This stems from a long-held misconception that the larger and more powerful of the sexes must be the male.)

Although in Arab eyes the female peregrines are essentially intended for the flight at *houbara*, they are also flown at *kairowan*. It is against this swift-flying, highly manoeuvrable quarry that the peregrine can really demonstrate her powers of flight. To find *kairowan* is by no means easy. They must be tracked in the desert, with the tracker travelling in or walking in front of the vehicle carrying the hawking-party. This is, by necessity, a four-wheel drive – frequently with an open top. While the hawks bump along in the back on a variety of makeshift perching points, the eyes of the tracker scan the ground, seeking the faint impression of a *kairowan*'s footstep.

When tracks are found, a falcon will frequently be unhooded to act as a spotter. When she bobs her head repeatedly in one direction, the hawking party can use binoculars to spot the *kairowan* which she has sighted. She is rehooded before the vehicle speeds towards the prey to flush it. When the quarry is airborne, the falcon is unhooded and leaves the fist. The *kairowan* will begin to climb, with the peregrine in hot pursuit. The *kairowan* will try to outfly her, twisting and turning every time she gets within striking distance. As in rook-hawking, the peregrine will try to get above the *kairowan*, so that she can stoop.

A *kairowan* which loses its nerve will drop to the ground to seek shelter. The peregrine will follow, aiming first to prevent it from taking off again and then to snatch it on the ground. At the last moment, if the quarry has strength left, it will leap into the air once more, jinking and turning for all it is worth to evade capture. The peregrine should stay right on its tail, aiming to demoralise it into making a fatal error of judgement, but the level of concentration sometimes takes its toll on the falcon and she will fly wide, causing the falconer to call her down to the lure.

The flights can end in a variety of ways. The *kairowan* may be taken on the ground. It may be struck in the air, whereupon all resistance will desert it and it will crumple up and fall to the ground, or it may be bound to and borne triumphantly down to earth. It may, however, prove to be the stronger and outfly the peregrine, if her stamina deserts her and, short of breath, she is forced to discontinue the chase and head ignominiously back to the falconer's lure.

The *tiba* or tiercels are the most exciting to watch during flights at *kairowan*, as their smaller size and tighter turning-circle results in a breathtaking contest of aerodynamics. The larger falcons, with their longer wings, lose more ground to the *kairowan* on the turn – a fact which is clearly visible from the ground, as the gap between them widens every time the quarry jinks. It requires all of a falcon's speed and persistence to overcome this disadvantage and achieve a kill.

Young *kairowan* are no match for a peregrine. When *kairowan* are found in groups, it is not unusual for an adult to feign weakness to attract the falcon's attention away from the younger birds. Adults will even sacrifice themselves on occasion. Female peregrines are capable of carrying *kairowan*, but tiercels are too small to do this.

*Houbara* are altogether a different prospect. Although hawks are flown at little bustard, which weigh about two and a half pounds, the species most frequently flown at is the Macqueen's bustard, the males of which average five and a half pounds in weight. While the sheer size of the *houbara* may make it appear superficially to be slow on the wing in comparison to the *kairowan*, this illusion is belied by its powerful flight and stamina. Peregrines are prized for flights at *houbara* because of their impressive speed and single-mindedness during a long chase. When the *houbara* is on the wing, a peregrine will follow it up very high, and she is extremely agile in pursuit. However, the very best *houbara* are capable of outringing and consequently outflying any falcon, whether it is peregrine, saker or gyrfalcon, so no chase can be considered to have a foregone conclusion.

A flight will be initiated in the same manner as those at *kairowan*, with tracking or spotting followed by the vehicle speeding towards the quarry. Once airborne, the *houbara* will either attempt to outfly the falcon or will land on the ground. A five-and-a-half pound *houbara* prepared to fight it out on the ground is no mean prospect for a falcon, which weighs only in the region of two pounds. Because of this, the hunting-party will often drive at a *houbara* which lands, forcing it into the air once more.

The flights can cover quite a distance as the two birds course across the skies, battling for supremacy. If the peregrine connects with the *houbara*, the two will often tumble to the ground together, or the hawk will strike and follow the prey down. On the ground, a *houbara* still in control of its faculties will reveal a particularly obnoxious defence mechanism, which it uses as a last resort. It will squirt at the falcon faeces of a particularly sticky variety called *'tamal'*. If its aim is good, the falcon will find that her feathers are gummed together and she will be unable to fly. If, however, the falcon is quick and bold enough, she can bind to the head of the *houbara*, hold it down, and bite through the neck to deliver the *coup de grâce*. Alternatively, she will subdue it and await assistance from the falconer.

When the falconer arrives on the scene, he will ensure that the *houbara* is dead, before covering the carcass in sand so that the peregrine, unable to see her prey, will release her hold on it and step up on to the glove or the *mangalah* – the protective cuff favoured in place of the glove in the Gulf states. The dark flesh of the *houbara* is very good to eat, and will be served with rice at evening camp.

Although falconry originated for a highly practical purpose – hunting for the pot – nowadays a falconer will freely admit that he and the vast majority of his fellow sportsmen fly peregrines purely for recreation and pleasure. However, the undisputed ability of the peregrine and her tractability when in expert hands have prompted her use for certain professional purposes. In Britain, the Fleet Air Arm

*Peregrine in Dubai on* wakr *– arabic perch –*
*with* mangalah *or padded cuff, used in place of*
*the falconry glove.*

trains its own men to fly peregrines in order to clear runways of other birds which might be sucked into the engines of planes during landing or take-off. Some United States Air Force bases contract professional falconers for the same purpose. Peregrines are also used in Europe to clear gulls from rubbish tips and flocks of birds from towns where they are causing damage to buildings. In the future, falconers will no doubt find other practical uses to which the peregrine can be put. Thus, the historic art of falconry is brought full circle as the peregrine once again serves the needs of man.

# THE FUTURE FOR THE PEREGRINE

*The peregrine falcon … has its legitimate and important place in the great scheme of things, and by its extinction, if that should ever come, the whole world would be impoverished and dulled.*

G. H. Thayer

The decline of the peregrine as a result of pesticides played a vital role in the future of a great many species, for she alerted a complacent populace to the dangers of their chemical world, imbuing some of them with a sense of responsibility. In the words of Tom Cade, 'the peregrine falcon has become a global symbol which has rallied a great diversity of human interests and activities to the cause of biological conservation'.

What horrors face the peregrine in the future? Is she safe now? Indeed, is it necessary even to speculate on her future prospects – could it not now be assumed that she has faced the worst and has recovered her numbers to a safe level? Sadly, mankind has such a poor track record with so many species on this planet that complacency would seem ill-advised.

The current status of the peregrine, viewed in her position as one of only four living bird species to achieve global distribution without the help of man, is considered to be 'threatened'. While she has repopulated many of her former haunts in the USA and parts of Europe, she is still endangered in certain parts of her range. In exploring her vulnerability, it is important to look at both her natural and her unnatural enemies if man is to take measures to ensure her survival.

In terms of natural enemies, one must consider the effects of mammalian predation, parasites and disease. The effects of the first of these appear to be minimal. Predominantly a cliff-nester, the peregrine has evolved to reduce the risks of predation of eyasses from the nest. Foxes are extremely wary of peregrines, who have been known to attack them vigorously when provoked within the immediate vicinity of their nest site. Smaller mammals, such as weasels, stoats, mink and polecats, have never been recorded as successfully raiding a peregrine eyrie. Apart from the effects of mammalian predation on elements of her food supply, there is therefore no evidence to suggest that any mammal is significant in the peregrine's population biology, except man.

There is little evidence that parasites cause high mortality in peregrines. As for disease, epidemics are unknown. However, studies in the 1960s showed that wild populations are subject to a handful of common problems. These principally include trichomoniasis or 'frounce', a protozoan causing canker of the mouth and throat, carried by pigeons; botulism, transmitted most commonly by the consumption of infected waterfowl; myiasis, maggot-fly infestation; and filaria, thread-worms which enter the blood supply. In addition, allowances must be made for death due to old age – impaired hunting ability due to decrepitude is an obvious killer of any predator. But, over and above these natural hazards, the biggest threat to the peregrine must come from man and his insidious pervasion of chemicals throughout the planet. The peregrine's sensitivity to pollutants has so nearly led to her extinction once. What other horrors does man have in store for her?

Environmental dangers are not limited to toxic chemicals: they also include industrial and urban development, intensification of land use, and 'advances' in agriculture. Not only does man poison the peregrine's environment, he also mounts specific attacks against her. Many gamekeepers, frequently with the approval of landowners, still shoot and poison her. Egg-collectors continue illicitly to steal her eggs, and taxidermists her skin. Prosecutions for 'wilful' disturbance at her nest sites are still often necessary. Only hawk-dealers have largely ceased to cause her problems, for she now breeds so readily in captivity that her vastly diminished 'commercial' value and ready accessibility render her not worth stealing from the wild.

In assessing the peregrine's future prospects, a good starting-point is to study her basic needs. She needs a stable food supply, suitable nesting and hunting territories, and clean air to breathe. Given the catholicism of the peregrine's tastes, a stable food supply should be reasonably unproblematic. It has been proven, however, that pairs will fail to reproduce in the same numbers as before if a favourite food source becomes less readily available. Evidence of this link between food availability and reproduction has been manifested in the past in relation to the western Scottish Highlands and the west of Ireland, where acidic rocks and poor soils, combined with intensive grazing, have reduced the land's capacity to support the wildlife on which the peregrine preys. In the future, similar problems are most likely to affect coastal peregrines, following any massive oil spillages decimating seabird populations.

Encroachment on the peregrine's nesting territories is likely to become an area of concern in the years ahead. In Britain, the forestry companies are planting the Scottish uplands at an alarming rate. In areas such as the western Highlands, where pigeons are scarce, the peregrines will suffer from any further afforestation of moorland. Indeed, any areas of dense forest planted in the future will cause the

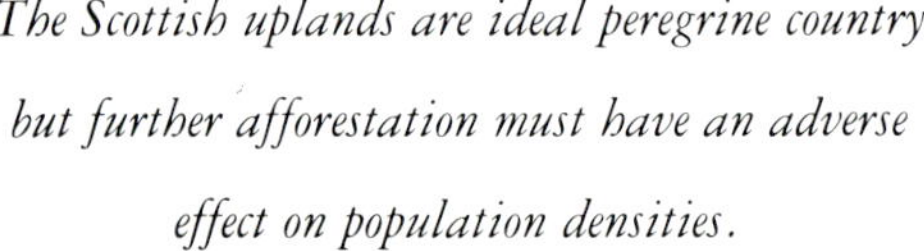

*The Scottish uplands are ideal peregrine country*
*but further afforestation must have an adverse*
*effect on population densities.*

peregrine problems, for unless there are wide tree breaks she will be unable to hunt. The draining and afforestation of open bogs in Finland had a highly detrimental effect on the Finnish ground-nesting peregrines, who thereby suffered a major loss of nesting habitat.

Areas of land which yield an abundance of food unsuited to the peregrine can cause her to be displaced by other raptors. This has been demonstrated by the increase in golden-eagle numbers in areas of intensive sheep-farming in Scotland, where the sheep carcasses provide easy pickings for the eagles, prompting the displacement of the peregrine from a significant number of her historical haunts.

The provision of clean air to breathe is plainly a problem which affects all life on this planet. As man continues to expel his poisons into the atmosphere, he must be aware that not all his new products can be tested conclusively for long-term safety. It is horribly likely that major mistakes will again be made – perhaps in the same league as the DDT problem, and possibly even worse – for the insidious nature of chemicals is such that they can tenaciously linger in the environment for many decades, affecting wildlife in ways which had never been anticipated. Indeed, DDT was popular specifically because it was so difficult to break down, and therefore its long-lasting effects as a pesticide made economic sense.

The only hope for the peregrine in the future is to continue with the monitoring of her derivatives and her environment. The eggs collected on the Big Sur coastline in California are still so full of toxins that they have to be handled with gloves. Residual DDT is still found in the blood of peregrines living in the wilderness of the Canadian Arctic and in their eggs, the shells of which are within one per cent of the 'thinning level'. The monitoring work which continues to be carried out relates to a crisis which is now decades old, but, without such studies, future crashes in the population may not be discovered until it is too late.

As global warming occurs, so the peregrine, who is less suited to an arid climate,

*Thin-shelled egg showing the damage caused by*
*DDT.*

is decreasing in certain areas in North America, where she is being replaced by the prairie falcon.

There are many areas in need of further and, in many cases, continual study. The mortality of wild populations is relatively easy to monitor owing to the peregrine's philopatry to her natal area. The captive populations also merit a great deal of work over the coming years, in the collating of biological data in order to determine whether pre-fledging characteristics or parentage relate to the future survival of released birds. The artificial manipulation of nest sites is a proven – though controversial – option open to scientists in areas where they need to boost the population.

There is still not much known about how the Third World peregrines are faring, although there have been many recent advances – pioneering work in the central

Andes of South America has revealed previously unknown breeding populations, for example. DDT is still being used in Third World countries, and will doubtless continue to be used until a safe, effective and economical alternative has been found. Research into an acceptable replacement pesticide is vital if peregrines and other raptors are to survive in the Third World. Information on the status of the peregrine in the Indian subcontinent and the Far East – and also in the former Soviet Union – is urgently needed. The workload does not diminish.

When pursuing a goal, it is essential to have a clear idea of exactly what that goal is. Should the peregrine merely recover to pre-DDT levels, or should the species be given a broader base? Through monitoring her progress, it has become apparent that populations left to recover naturally will most easily repopulate areas which lie adjacent to a remnant population. Regions from which the peregrine has been totally

extirpated, such as the south-east of England, will take much longer to recover. However, the peregrine *is* a survivor, and she *is* recovering her territories – albeit slowly. The remaining fear is for the gene pool in certain regions – for example, that in the eastern North American states no longer resembles the pre-DDT gene pool of *Falco peregrinus anatum*, for captive releases have created a novel regional gene pool, built up from *anatum*, *tundrius*, *brookei*, *pealei*, *cassini* and other subspecies.

While monitoring and research will play a critical role in the peregrine's continued survival, it is essential that these are complemented by education. If people are not aware of the peregrine in all her glory, why should they bring pressure to bear to ensure her future? If she is to endure, each successive generation must champion her cause. To this end, peregrines held in captivity can be used to educate children at special centres, in lectures to schools, and in demonstrations of her peerless mastery of the air. Films, books and television can be utilised to broadcast her image as widely as possible in an effort to increase public support. Man can be justifiably proud of the great strides he has made within the field of global communications; the perceptive use of this communications network to inspire public support for threatened wildlife is vital.

Those who raise their heads to salute this magnificent bird as she cleaves our skies above must surely suspect that her continued survival or demise will prefigure our own. The peregrine is gifted with such perfection of physical form, she enjoys extraordinary hunting ability, and she is endowed with the rare adaptability which enables her to thrive both in the wilderness and cheek by jowl with man in his urban environment. If she cannot endure in this world, what hope is there for us?

## PEREGRINE SUBSPECIES

Working on the '75 per cent rule' – the idea that certain recognisable traits are present in 75 per cent of a given population of a particular subspecies – the following are the most commonly deemed to be peregrine subspecies. They are listed in approximate order of size, starting with the largest: it should be stressed that the guides to colouration given below are generalisations. Individuals from within a race can show a great deal of variation, both in adult and immature plumage.

*FALCO PEREGRINUS PEALEI* (PEALE'S FALCON, PEALE'S PEREGRINE)

Peale's falcons breed in North America, on the North Kurile Islands and Aleutians, south to the Queen Charlotte Islands. They are dark slaty-blue above, with moderately heavy wide black barring on the breast and some spots. The ground colour of the chest feathers is bluish-white. The feet, cere (waxy skin above the beak) and lore (skin around the eye) are pale lemon-yellow, rather than the bright orange-yellow more typical of other subspecies. The weight of the tiercels ranges from 1lb 13oz to 2lb 5oz, and of the falcons from 2lb 12oz to 3lb 8oz. In build they are long, broad, powerful birds, with wide tails.

*FALCO PEREGRINUS CALIDUS* (SIBERIAN PEREGRINE, EURASIAN TUNDRA PEREGRINE)

*Calidus* occur in Eurasia, as far north as seventy-six degrees north in Russia, Siberia and Lapponia and as far east as the Lena. They migrate as far south as Natal and New Guinea. The back feathers are predominantly blue, with a white, sparsely spotted chest. The cheeks are white, and the moustachial stripe is not wide. The tiercels weigh from 1lb 5oz to 1lb 10oz and the falcons from 1lb 13oz to 2lb 15oz.

*FALCO PEREGRINUS JAPONENSIS* (JAPANESE PEREGRINE)

Found in East Asia, *japonensis* occur with *calidus* in eastern Siberia and breed as far east as the Kurile Islands. They winter in Japan, Riu Kiu and Taiwan. In colouration they are black on the back and blue-buff on the front, fading to grey at the sides. On the chest, they have wide and fairly heavy bars, with spots on the crop. The only recorded weight is from one tiercel which weighed 1lb 2oz.

*FALCO PEREGRINUS ANATUM* (DUCK HAWK, GREAT-FOOTED HAWK, AMERICAN PEREGRINE FALCON, ROCK PEREGRINE)

*Anatum* breed in North America – in Labrador, eastern Greenland, Alaska, and as far south as southern California. They are blue on the back and rust on the front, with a medium- to narrow-barred chest, sparingly spotted. The moustachial stripe is wide. The tiercels weigh from 1lb 2oz to 1lb 9oz and the falcons from 1lb 14oz to 2lb 11oz.

*FALCO PEREGRINUS PEREGRINUS* (PEREGRINE FALCON, EUROPEAN PEREGRINE)

Spread across Eurasia, *F. peregrinus peregrinus* occur east to North Russia and south to the Mediterranean and the Caucasus. Typically, they are blue on the back and whitish-buff on the front, but they can be pinkish on the chest. They have medium,

moderately dense barring on the chest, with some spots. Their weight ranges from tiercels of 1lb 3oz to 1lb 7oz up to falcons of 1lb 10oz to 2lb 7oz.


*FALCO PEREGRINUS FURUITII* (VOLCANO ISLANDS PEREGRINE, IWO PEREGRINE)

*Furuitii* breed on the Volcano Islands, south-east of Japan, and on the Bonin Islands. They are black on the back and dark buff on the chest. The chest is heavily barred, with some spots, and the tail is black. The moustache is of medium width. No weights have been recorded.


*FALCO PEREGRINUS MADENS* (CAPE VERDE ISLANDS PEREGRINE)

*Madens* breed only on the Cape Verde Islands. A little smaller than *falco peregrinus peregrinus*, they are much browner, with a rust brown on the crown, nape and collar. The barring on the front is of medium width and spacing, on a tawny to buff-washed chest. The back is brownish-blue. No weights have been recorded.


*FALCO PEREGRINUS TUNDRIUS* (TUNDRA FALCON, TUNDRA PEREGRINE, ARCTIC PEREGRINE, BEACH PEREGRINE)

*Tundrius* occur in North America, from the Bering Strait to northern Baffin Island. Their southern range extends to approximately sixty-five degrees north. They were identified as a subspecies by Dr Clayton M. White in 1968. With a predominantly blue back and a white front, *tundrius* are a pale race, with a light head and few bars on the chest. They frequently have a predominance of spots on the crop and in place of bars on the middle parts of the chest. *Tundrius* are highly migratory, crossing the Equator to winter in southern Brazil and northern Argentina. Recorded weights list tiercels of 1lb 2oz to 1lb 10oz and falcons of 1lb 11oz to 2lb 6oz.

Falco peregrinus
pelegrinoides

Falco peregrinus
peregrinator

Falco peregrinus
peregrinus

Falco peregrinus
pealei

*FALCO PEREGRINUS CASSINI* (CASSIN'S FALCON, AUSTRAL PEREGRINE)

*Cassini* breed in South America, in Chile, from Atacama south to Tierra del Fuego, in southern Patagonia and on the Falkland Islands. This subspecies exists in both a pale and a dark morph, or form, the difference manifesting itself in the immature plumage and perpetuated in the adult plumage. Typically, in the dark morph the head and sides of the head are completely black and the chest is buff, with bars and spots. The back is blue-grey, with black bars, particularly on the wing-coverts. The pale morph has a light-grey back and a light-buff chest, with no bars whatsoever, but with light brownish-grey streaks on the crop and abdomen, changing to arrowheads on the thighs. The head is very pale. This pale-colour morph is now considered to be what was formerly known as *Falco kreyenborgi* – Kleinschmidt's falcon or the pallid falcon. Both dark and pale morphs have been found in the same nest, but the pale morph is mainly found in southern Patagonia and Tierra del Fuego and is considered to be extremely rare. No weights are available.

*FALCO PEREGRINUS BROOKEI* (LESSER PEREGRINE, MEDITERRANEAN PEREGRINE, SPANISH PEREGRINE)

*Brookei* are found in the Mediterranean, particularly on the coast, and in Asia Minor, from southern Spain and northern Morocco, east to the Caucasus. They are non-migratory. Their colouration is pinkish on the chest with heavy barring. The nape is bright rufous, and the back is dark blue to black. There is a recorded weight for a tiercel of 1lb and two falcons have been noted at 2lb.

*FALCO PEREGRINUS BABYLONICUS* (RED SHAHEEN, RED-NAPED SHAHEEN)

*Babylonicus* occur in the central deserts and steppes of Asia, from Iraq and eastern Iran, east to Mongolia. They winter partly in India. They are the palest of all the races, with a light blue-brown back and a buff belly, sparingly barred. The crown and the nape are rufous, similar to those of the lanner falcon. The palest individuals are from Iran, Baluchistan and Afghanistan, while darker birds occur in eastern Turkistan and western Mongolia. In weight, the tiercels average 12 to 14 oz, and the falcons 1lb 2oz to 1lb 11oz.

Although *babylonicus* is considered by some to be a separate species, together with *pelegrinoides*, the Barbary falcon, the two are definitely more closely related to the peregrine than to any other species of falcon. In addition, there is one case of interbreeding on record between *babylonicus* and *peregrinator*, the black shaheen. Thus, subject to further studies, the majority of scientists consider the Barbary and the red-naped shaheen to be short-tailed desert races of the peregrine.

### *FALCO PEREGRINUS NESIOTES* (ISLAND PEREGRINE, FIJI PEREGRINE)

*Nesiotes* literally means 'the islander'. They are found in the south-west Pacific islands of New Hebrides, New Caledonia, Tanna, Loyalty and Beaupré. It has not yet been firmly established whether the peregrines found on Fiji are also *nesiotes*. The overall colouration of *nesiotes* is very dark, with head and cheeks of solid black, causing the moustachial stripe to become a triangle. The feathering of the back is black, edged with grey. On the chest they have medium to narrow barring, over a rust wash, tinged with grey on the thighs and flanks. No weights have been recorded.

### *FALCO PEREGRINUS ERNESTI* (ERNEST'S FALCON, HOSE'S FALCON)

*Ernesti* are found in Indonesia, the Philippines, New Guinea and neighbouring islands. They are reputed to be the darkest of all the peregrine subspecies. They are virtually completely black on the back, with tiny grey fringes to the feathers. The head is totally black-capped, like that of *nesiotes*, and the chest is dark bluish-grey,

with rufous overtones. The thighs are blackish, with very heavy barring, which is also present on the under-wing coverts. No weights have been recorded.

*FALCO PEREGRINUS MACROPUS* (BLACK-CHEEKED FALCON, AUSTRALIAN PEREGRINE, DUCK HAWK, PIGEON HAWK)

*Macropus* is one of two subspecies found in Australia. They are found throughout the continent, except in the south-west. They also breed on Tasmania. They are black or dark blue on the back, with some variegation. The chest is rusty buff, densely barred. Two individuals have had their weights recorded – a tiercel at 15oz and a falcon at 2lb 2oz.

*FALCO PEREGRINUS SUBMELANOGENYS* (BLACK-CHEEKED FALCON)

*Submelanogenys* are the other Australian peregrines. They occur in the south-west, where *macropus* is absent. They are similar to *macropus* on the back and head, but the whole underside is rufous – pale on the throat and crop, but very dark on the abdomen and legs, with dense black barring. No weights have been recorded.

*FALCO PEREGRINUS MINOR* (AFRICAN PEREGRINE, LESSER PEREGRINE)

*Minor* are found in Africa, south of the Sahara, from Ghana to northern Ethiopia and south to Cape Province. The head is black, and the moustachial stripe is very wide. The back is black with blue edges to the feathers. The underside of the throat and upper crop are white, fading to buff on the lower crop and abdomen. There can be traces of rufous in the nape and crown. In weight, the tiercels average 1lb and the falcons 1lb 8oz.

*FALCO PEREGRINUS PEREGRINATOR* (BLACK SHAHEEN, INDIAN PEREGRINE)

Black shaheen range from India and eastern Sri Lanka to southern China. They are

dark grey above, with a deep orange on the lower crop, abdomen and flanks. The barring varies from heavy to medium, with the crop either lightly streaked or clear of markings. The tail is tinged with rufous, and in the darkest individuals from India the orange colouration extends right up to the cheeks. The head is very black. One tiercel has been weighed at 1lb 4oz.

*FALCO PEREGRINUS RADAMA* (MADAGASCAR PEREGRINE)
*Radama* breed on Madagascar and the Comoro Islands. They are dark blue-grey above, with white chests, black barring and spots. No weights have been recorded.

*FALCO PEREGRINUS PELEGRINOIDES* (BARBARY FALCON)
The smallest of all the peregrines, Barbaries have comparatively long wings and rounded tails. They are found in southern Egypt and northern Sudan, Arabia and North Africa, as far west as the southern Atlas and the Atlantic coast in southern Morocco. They are light blue-brown on the back, with a rufous head and a thin moustachial stripe. On the chest they are white to buff, with narrow bars or spots, and arrowheads on the thighs. Tiercels weigh from 12 to 13 oz and falcons from 1lb 6oz to 1lb 9oz.

# BIBLIOGRAPHY


Allen, Mark, *Falconry in Arabia* (London: Orbis, 1980)

Audubon, John James, *Birds of America*, facsimile ed. R. T. Peterson and V. M. Peterson (New York: Abbeville, 1990)

Ayala, Pero Lopez de, *Libro de la Caza de las Aves*, ed. John Cummins (London: Tamesis Books, 1987)

Baker, J. A., *The Peregrine* (London: Collins, 1967)

Bannerman, D. A., *Birds of the British Isles* (Edinburgh: Oliver & Boyd, vol.1 1953, vol. 2 1956)

Barnes, Dame Julians, *see* Hands, Rachel

Bayer, William, *Peregrine* (London: Severn House, 1982)

Beebe, Frank Lyman, and Webster, Harold Melvin, *North American Falconry and Hunting Hawks* (Denver, Colorado: North American Falconry and Hunting Hawks, 5th edn, 1985)

Blaine, Gilbert, *Falconry* (London: Philip Alan, 1936; Spearman, 1970)

Bodio, Stephen, *A Rage for Falcons* (New York: Schocken Books, 1984)

Brown, Leslie, and Amadon, Dean, *Eagles, Hawks and Falcons of the World* (Feltham: Country Life Books, 1968)

Cade, Tom J., *The Falcons of the World* (London: Collins, 1982)

Cade, Tom J., Enderson, James H., Thelander, Carl G., and White, Clayton M., *Peregrine Falcon Populations, Their Management and Recovery* (Boise, Idaho: Peregrine Fund, 1988)

Carson, Rachel, *Silent Spring* (New York: Crest Books, 1962)

Chamerlat, Christian Antoine de, *Falconry and Art* (London: Sotheby's, 1987)

Cummins, John, *The Hound and the Hawk: The Art of Medieval Hunting* (London: Weidenfeld & Nicolson, 1988)

Dennis, Roy, *Peregrine Falcons* (Lanark: Colin Baxter Photography, 1991)

Ford, Emma, *Bird of Prey* (London: Batsford, 1982)

—, *Falconry in Mews and Field* (London: Batsford, 1982)

—, *Falconry* (Princes Risborough: Shire, 1984)

—, *Falconry, Art and Practice* (London: Blandford, 1992)

Frederick II, Emperor of the Holy Roman Empire, C13th manuscript in Vatican Library; 1317 copy in Bibliothèque Nationale de Paris; several translations exist including *The Art of Falconry, being the 'De Arte Venandi cum Avibus' of Frederick II of Hohenstaufen*, trans. and ed. by Casey A. Wood and F. Majorie Fyfe (Stanford, Ca: Stanford University Press, 1943)

Gaski, Andrea L., Jenkins, M. Allen, and Porter, Richard D., *The Working Bibliography of the Peregrine* (Washington DC: Institute for Wildlife Research, 1987)

Gould, John, *Birds of Britain* (London, 1873)

—, *Birds of Europe*, (London, 1837)

Grossman, Mary Louise, *Birds of Prey of the World* (London: Cassell, 1965)

Hagar, J. A., 'History of the Massachusetts Peregrine Falcon Population 1935–57' in J. J. Hickey (ed.), *Peregrine Falcon Populations, Their Biology and Decline* (Maddison: University of Wisconsin Press, 1969)

Hands, Rachel, (ed.), *English Hawking and Hunting in 'The Boke of St Albans'* (Oxford: Oxford University Press, 1975)

Hangte, E., '*Zum Beuterwerb unserer Wanderfalke*' (*Ornithol Mitt* 20: 211–217, 1968)

Harting, James Edward, *The Ornithology of Shakespeare: Critically Examined, Explained and Illustrated* (Old Woking: Gresham Books, 1978)

Heatherley, Francis, *The Peregrine Falcon at the Eyrie* (London: Country Life Books, 1913)

Jameson, E. W., Jr, *The Hawking of Japan* (Davis, Ca: University of California, 1962)

Lodge, George E., *Memoirs of an Artist Naturalist* (London and Edinburgh: Gurney & Jackson, 1946)

Lydekker, R., *Royal Natural History*, 6 vols. (London: Warne, 1893–6)

Massinger, Philip, *The Guardian* (1633) quoted in Harting, *op. cit.*

Mavrogordato, Jack, *A Falcon in the Field: A Treatise on the Training and Flying of Falcons* (London: Knightly Vernon, 1966)

Meinertzhagen, R., *Pirates and Predators: The Piratical and Predatory Habits of Birds* (Edinburgh and London: Oliver and Boyd, 1959)

Peterson, Roger Tory, *Birds over America* (1942)

Ratcliffe, Derek, *The Peregrine Falcon* (Calton, Staffs: Poyser, 1980)

Ryves, B. H., *Bird Life of Cornwall* (London: Collins, 1948)

Salvin, and Broderick, *Falconry in the British Isles* (London: John van Voorst, 1855)

Savage, Candace, *Peregrine Falcons* (London: Robert Hale, 1992)

Schlegel, Hermann, and Wulverhorst, A. H. Verster de, *Traité de Fauconnerie* (London: Pion, 1979)

Shakespeare, William, *The Complete Works*, ed. Stanley Wells and Gary Taylor (Oxford: Oxford University Press, 1986)

Sherrod, Steve K., *Behaviour of Fledgling Peregrines* (Fort Collins, Colo: Pioneer Impressions, 1983)

Spenser, Edmund, *The Faerie Queen*, in *Spenser: Poetical Works*, ed. J. C. Smith and E. de Selincourt (Oxford: Oxford University Press, 1912)

Swann, H. Kirke, *Monograph of the Birds of Prey* (London: Wheldon & Wesley Ltd, 1930)

Thorburn, A., *Birds of Prey* (London: W. F. Embleton, 1919)

Treleaven, R. B. *Peregrine: The Private Life of the Peregrine Falcon* (Penzance: Headland Publishing, 1977)

Turbervile, George, *The Booke of Faulconrie or Hawking* (1575)

Upton, Roger, *A Bird in the Hand: Celebrated Falconers of the Past* (London: Debrett, 1980)

Upton, Roger, *O for a Falconer's Voice* (Marlborough, Wilts: Crowood, 1987)

Waller, Renz, *Der Wilde Falke ist mein Geselle* (Berlin: Neumann-Neudamm, 1982)

Walton, Isaak, *The Compleat Angler* (London: Rich. Marriot, 1653)

Webster, Hal, and Enderson, James, *Game Hawking at its Very Best* (Denver, Colo: Windsong Press, 1988)

Weick, Friedhelm, and Brown, Leslie H., *Birds of Prey of the World* (London: Collins, 1980)

Williamson, Henry, *The Peregrine's Saga and Other Wild Tales* (London: Putnam, 1945)

Zaid bin Sultan Al Nahayan, *Falconry as Sport* (Westerham: Westerham Press, 1976)

Emma Ford has been a falconer since the age of eight. With her husband, Steve, she founded the internationally acclaimed British School of Falconry in 1982. The two branches of the School, at the Gleneagles Hotel and on the Braco Castle Estate, in Perthshire, Scotland attract enthusiasts from all over the world.

Emma has travelled widely in connection with her work, teaching, acting as a consultant and presenting papers at international zoological and veterinary conferences. Former Countrywoman of the Year, she drew up the falconry syllabus for the Duke of Edinburgh Award scheme and is an elected member of the Hawk Board – the government advisory panel on captive hawks. Her international reputation is highlighted by frequent television appearances, documentaries and articles outlining her work in national newspapers and magazines.

A keen conservationist and country-sportswoman, Emma writes for national newspapers and magazines. She wrote her first book at the age of sixteen; this is her fifth published title.